CHEMICAL ANALYSIS

Vol. 1. **The Analytical Chemistry of Industrial Poisons, Hazards, and Solvents.** *Second Edition.* By Morris B. Jacobs
Vol. 2. **Chromatographic Adsorption Analysis.** By Harold H. Strain (*out of print*)
Vol. 3. **Colorimetric Determination of Traces of Metals.** *Third Edition.* By E. B. Sandell
Vol. 4. **Organic Reagents Used in Gravimetric and Volumetric Analysis.** By John F. Flagg (*out of print*)
Vol. 5. **Aquametry: Application of the Karl Fischer Reagent to Quantitative Analyses Involving Water.** By John Mitchell, Jr. and Donald Milton Smith (*temporarily out of print*)
Vol. 6. **Analysis of Insecticides and Acaricides.** By Francis A. Gunther and Roger C. Blinn (*out of print*)
Vol. 7. **Chemical Analysis of Industrial Solvents.** By Morris B. Jacobs and Leopold Scheflan
Vol. 8. **Colorimetric Determination of Nonmetals.** Edited by David F. Boltz
Vol. 9. **Analytical Chemistry of Titanium Metals and Compounds.** By Maurice Codell
Vol. 10. **The Chemical Analysis of Air Pollutants.** By Morris B. Jacobs
Vol. 11. **X-Ray Spectrochemical Analysis.** By L. S. Birks
Vol. 12. **Systematic Analysis of Surface-Active Agents.** By Milton J. Rosen and Henry A. Goldsmith
Vol. 13. **Alternating Current Polarography and Tensammetry.** By B. Breyer and H. H. Bauer
Vol. 14. **Flame Photometry.** By R. Herrmann and J. Alkemade
Vol. 15. **The Titration of Organic Compounds** (*in two parts*). By M. R. F. Ashworth
Vol. 16. **Complexation in Analytical Chemistry: A Guide for the Critical Selection of Analytical Methods Based on Complexation Reactions.** By Anders Ringbom
Vol. 17. **Electron Probe Microanalysis.** By L. S. Birks
Vol. 18. **Organic Complexing Reagents: Structure, Behavior, and Application to Inorganic Analysis.** By D. D. Perrin
Vol. 19. **Thermal Methods of Analysis.** By Wesley Wm. Wendlandt
Vol. 20. **Amperometric Titrations.** By John T. Stock
Vol. 21. **Reflectance Spectroscopy.** By Wesley Wm. Wendlandt and Harry G. Hecht
Vol. 22. **The Analytical Toxicology of Industrial Inorganic Poisons.** By the late Morris B. Jacobs
Vol. 23. **The Formation and Properties of Precipitates.** By Alan G. Walton
Vol. 24. **Kinetics in Analytical Chemistry.** By Harry B. Mark, Jr. and Garry A. Rechnitz
Vol. 25. **Atomic Absorption Spectroscopy.** By Walter Slavin
Vol. 26. **Characterization of Organometallic Compounds** (*in two parts*). Edited by Minoru Tsutsui
Vol. 27. **Rock and Mineral Analysis.** By John A. Maxwell
Vol. 28. **The Analytical Chemistry of Nitrogen and Its Compounds** (*in two parts*). Edited by C. A. Streuli and Philip R. Averell
Vol. 29. **The Analytical Chemistry of Sulfur and Its Compounds** (*in two parts*). By J. H. Karchmer
Vol. 30. **Ultramicro Elemental Analysis.** By Günther Tölg
Vol. 31. **Photometric Organic Analysis** (*in two parts*). By Eugene Sawicki
Vol. 32. **Determination of Organic Compounds: Methods and Procedures.** By Frederick T. Weiss
Vol. 33. **Masking and Demasking of Chemical Reactions.** By D. D. Perrin

CHEMICAL ANALYSIS

A SERIES OF MONOGRAPHS ON

ANALYTICAL CHEMISTRY AND ITS APPLICATIONS

VOLUME 33

WILEY-INTERSCIENCE

Division of John Wiley & Sons, New York/London/Sydney/Toronto

Masking and Demasking of Chemical Reactions

THEORETICAL ASPECTS AND PRACTICAL APPLICATIONS

D. D. Perrin

Department of Medical Chemistry
Institute of Advanced Studies
Australian National University, Canberra

WILEY-INTERSCIENCE

A Division of John Wiley & Sons New York • London • Sydney • Toronto

Library of Congress Catalog Card Number: 79-114918

ISBN 0-471-68071-0

Printed in the United States of America

10 9 8 7 6 5 4 3 2 1

PREFACE

The concept of masking was introduced into chemistry by Professor Fritz Feigl to describe a process by which an ion or a molecule could be so transformed, usually by complex formation, that it no longer gave its typical chemical reactions. Subsequently masking (and demasking) techniques found widespread applications in many fields of chemistry, both pure and applied, because they provide a convenient means by which to remove, or temporarily suppress, the effects of unwanted constituents of a system without the necessity of recourse to physical separation.

In analytical chemistry masking is used to improve selectivity in chemical methods ranging over fields as diverse as complexometric titration, spectrophotometry, gravimetry, polarography, solvent extraction, and ion exchange. Industrial applications are similarly widespread, the main applications of masking procedures being to keep within acceptable limits the levels of free metal ions which could otherwise cause turbidities or precipitates, rancidity, decomposition, oxidation, color development, or other undesirable changes. Reagents able to mask metal ions in the human body are used in methods for the treatment of metal poisoning and for the removal of excessive stores of iron and copper from the body. They are also important in enzymology and in the preparation of culture media.

At the same time there has been considerable development in our understanding of chemical reactions and in quantitative information about equilibria in solution. This has made it possible to provide a basis on which to decide on likely masking agents for particular purposes. As Professor Anders Ringbom showed in an earlier book in this series, a treatment now exists for deciding on the most favorable conditions (of pH, reagent concentrations, types of masking agent, etc,) for an analysis and for assessing its relative accuracy. This approach, in turn, is readily adaptable to computer application, thereby providing a convenient means of assessing possible new analytical procedures, including cases in which interfering side reactions, solvent extractions, or precipitations are involved.

New analytical techniques such as neutron activation and atomic absorption spectrometry are diminishing the areas of conventional chemical analysis. The extent of this erosion will depend in large measure on how much can be achieved in simplifying and speeding up existing titrimetric, spectrophotometric, and gravimetric procedures without, at the same time, any concomitant loss of accuracy or reliability. It is in this field that much can be hoped for from simple and elegant masking and demasking techniques designed to enhance the selectivity of chemical reactions and in many cases to make possible the rapid analysis of mixtures that could not otherwise be dealt with easily.

Although a number of books have been written on organic sequestering agents and the use of complexometric methods in analysis, masking has not hitherto been separately dealt with in spite of the extensive but widely scattered references to it in the literature. The aim of this monograph is to provide, on the one hand, many representative examples of applications of masking and demasking in analysis, industry, and medicine and, on the other, some indication of the underlying principles to serve as a guide to the reader in his choice of masking reagents for use in new systems or under conditions not previously considered. It must not be assumed that the procedures described are necessarily the best for analytical purposes. In some cases alternative methods would undoubtedly be preferable.

Among the many contributors to the current knowledge of masking procedures special mention should be made of the debt we owe to Fritz Feigl who pioneered the field, R. Pribil and H. Flaschka who have labored mightily to improve complexometric titrations, and Gerold Schwarzenbach and Anders Ringbom who have pointed the way to the quantitative treatment of complexation reactions.

Canberra, Australia D. D. Perrin
April 1970

CONTENTS

CHAPTER

ABBREVIATIONS

BAL	2,3-dimercapto-1-propanol
CDTA	cyclohexane-1,2-diamine-N,N,N′,N′-tetraacetic acid
DDC	diethyldithiocarbamate
DTPA	diethylenetriaminepentaacetic acid
EDTA	ethylenediamine-*N,N,N′,N′*-tetraacetic acid
EGTA	ethyleneglycol-bis(2-aminoethylether)tetraacetic acid
ferron	7-iodo-8-hydroxyquinoline-5-sulfonic acid
HEDTA	2-hydroxyethylethylenediaminetriacetic acid
neocuproine	2,9-dimethyl-1,10-phenanthroline
PAN	1-(2-pyridylazo)-2-naphthol
PAR	4-(2-pyridylazo)resorcinol
1,10-phen	1,10-phenanthroline
tiron	disodium pyrocatechol-3,5-disulfonate
TTHA	triethylenetetramine-N,N,N′,N′,N″,N″-hexaacetic acid
unithiol	sodium 2,3-dimercaptopropane sulfonate

CHAPTER

1

INTRODUCTION

The concept of masking, in the sense of concealing or disguising, is a familiar one,[1] and so also is the idea of demasking in which this concealment or disguise is removed. Extension of these concepts to chemistry owes a great deal to Professor F. Feigl, whose monograph, *Specific and Special Reactions*,[2] made a major contribution to the understanding of the principles and applications of these techniques to analytical chemistry. In 1936 Feigl[3] defined *masking reagent* and the *masking* of a reaction in these terms: ". . . the concentration of a given ion in solution may be so diminished by the addition of substances which unite with the ion to form complex salts that an ion product sufficient to form a precipitate or cause a color reaction is no longer obtained. Thus we speak of the *masking* of a reaction, and call the reagent responsible for the disappearance of the ions necessary for the reaction, the *masking reagent*. In this way, and without having to resort to physical separations, Feigl was able to make spot tests highly selective or even specific.

The use of the word masking in chemistry has since become recognized in world literature. Thus *Der Grosse Brockhaus*[4] includes under *maskieren: "analytische Chemie:* Verhindern des normalen Reaktionsablaufs. Durch geeignete Komplexbildung wird die Konzentration einer Ionart so weit herabgesetzt, dass bei Zugabe eines charakterist. Reagens die sonst entstehende Reaktion unterbleibt. Durch M. können auftretende Störungen beseitigt und dadurch unspezifische Reaktionen zu eindeutigen Nachweisen verwendt werden."

Unfortunately *Webster's International Dictionary* (1934) gave a rather different definition of masking, namely "to mask (chem.): to prevent (an atom or group of atoms) from showing its ordinary reactions; as to mask hydroxyl in a sugar by converting it into methoxyl." This extension of the meaning of masking to include the use of blocking groups to protect part of an organic molecule during subsequent chemical operations has not become common usage and will not be followed here.[5]

A masking reagent, then, is one that lowers the concentration of a free metal ion or a free ligand to such a level that certain of its chemical reactions are prevented. Thus the addition of excess ammonia to Ni(II),

Cu(II), Zn(II), or Cd(II) will prevent the precipitation of the corresponding hydroxide from alkaline solutions. Similarly, in the presence of excess EDTA (ethylenediaminetetraacetic acid), Fe(III) solutions are unable to form red thiocyanate complexes. Again, many of the reactions of cyanide ion are prevented if mercuric ion is added to produce undissociated $Hg(CN)_2$.

In many instances only a weak masking effect is required, leaving the metal species free to react with more powerful complexing or reducing agents. Although ammonia prevents the precipitation of zinc as its hydroxide, it does not interfere in the titration of zinc with EDTA. Similarly, silver is masked by ammonia against precipitation as the hydroxide or chloride, but not against precipitation as the iodide. Complex formation with tartrate ion prevents the precipitation of copper(II) as its hydroxide, but a sufficient equilibrium concentration of free copper(II) remains so that certain sugars and other reducing agents can still be detected by the reduction of copper(II) to copper(I) oxide: this is the basis of Fehling's solution. Nylander's solution operates on a similar principle, except that bismuth, instead of copper, is masked by tartrate.

Masking agents are not specific for individual ions but most of them form stable complexes with only limited groups of metal ions. Examples include cyanide ion with Cu, Ag, Au, Zn, Cd, Hg, Fe, Co, and Ni, and fluoride ion with Mg, Ca, Al, other tri- and tetravalent cations. EDTA (which is usually employed as its soluble disodium salt) has little selectivity, the extent to which it reacts with a wide range of cations depending on the pH of the solution.

Varying the pH of a solution is one of the most important ways of changing the effectiveness with which a metal ion is masked by a complexing species. At high enough pH values most metal ions form hydroxy complexes which may be soluble, like $Al(OH)_4^-$ or $Zn(OH)_4^{2-}$, or insoluble, like $Mg(OH)_2$, so that hydroxyl ion, itself, can sometimes be an effective masking agent. More commonly, however, the masking agent is an organic acid or base which undergoes protonation as the pH of the solution is lowered, thereby diminishing, or even abolishing, its masking action. At pH 11, an $0.1M$ copper(II) solution contains less than $10^{-13}M$ free copper(II) ions if $1M$ excess of ammonia is present, whereas if the solution is brought to pH 3, less than 0.1% of the copper(II) is present as an ammine complex. A familiar example of pH-dependent complexing ability is in the use of hydrogen sulfide as a precipitant in the group separation of metals. Because H_2S is a weak acid protons compete strongly with metal ions for addition to the sulfide ion, S^{2-}, so that at lower pH values the only metals that can be precipitated are those forming very insoluble sulfides. With increasing pH of the solution proton competition becomes less important and metal ions forming less insoluble sulfides are also precipitated.

Because most ligands are proton acceptors their masking ability varies with pH in a way that depends specifically on their pK_a values, decreasing as the solution is made more acid. Also, in general, the stability constants of two different kinds of metal ions will not vary by a constant amount from one kind of complexing species to another. These two factors have important consequences for masking and demasking because, in most cases, use is made of these differences to form a complex between one kind of metal ion and a reagent (the *principal reaction*), whereas because of their more favorable equilibria with the masking agent, the reactivity of other metal ions is suppressed.

Masking procedures are usually of the following types:

1. A masking agent is added to a solution prior to a determination so as to form a stronger complex with the interfering ion than with the desired species which, instead, is able to react normally. Often, however, the masking agent is added to the metal–ion mixture after the total metal ion content has been determined by complexometric titration. The amount of metal ion displaced in this way by the masking agent can then be determined from the amount of free ligand that is liberated. An example is the addition of fluoride ion to a solution containing aluminum and other metal ions after their titration with EDTA, thereby converting the aluminum–EDTA complex to an aluminum–fluoride complex and free EDTA.

2. A masking agent changes the valence state of the interfering ion; for example, in acid solutions ascorbic acid reduces iron (III) to iron(II), and mercury(II) is reduced to the metal. Cysteine reduces copper(II) to copper(I). Alkaline peroxide oxidizes chromium(III) to chromate.

3. A masking agent precipitates the interfering ion in a form which can be left in the system without causing trouble in further determinations. Above pH 12 calcium can be titrated with EDTA in the presence of magnesium ion which is precipitated as the hydroxide.

4. Advantage is taken of differences in the rates of formation or dissociation of complexes by the ion to be masked and the ion to be determined. In the cold iron(III) can be titrated with EDTA in the presence of Cr(III) because the latter reacts only very slowly. Conversely, when bismuth(III) ion is added to an ice-cold solution at pH 2 containing EDTA complexes of Cd(II), Co(II), Cu(II), Pb(II), Ni(II), and Zn(II), nickel, unlike all the other cations, is only very slowly displaced and can be determined by difference. In addition, there are more specialized techniques, such as electrochemical masking used in polarography,

and kinetic masking in which advantage is taken of the inhibitory effect of metal-complex formation on reactions catalyzed by metal ions or inorganic anions.

The reverse procedure, demasking, occurs when a substance is added to counteract the effect of a masking agent which is already present in the solution; for example, fluoride ion is a masking agent for stannic tin against precipitation as stannic sulfide, but the tin is demasked and can be precipitated in this way, if boric acid is added to remove the fluoride ion by converting it to the very stable species BF_4^-.

The term *sequestering,* which is a common synonym for masking in cases in which the unwanted species is masked in a solution by a complexing agent, to form a soluble complex, was first used in this way to describe the softening of water by the use of sodium hexametaphosphate.[6] It is interesting to note, however, that Boyle used *sequester* in the sense of "to set apart, put aside, separate, segregate" when he wrote in *The Skeptical Chymist* (1661) that ". . ashes . . consist of pure salt and simple earth, sequestered from all the other principles or elements." Currently, sequestration might be defined as the diminution of the concentration of a free multivalent cation, by complex formation, to such a level that it is no longer precipitated by a given anion or otherwise exerts its customary chemical properties or reactions. This suppression of activity must be achieved without physical removal of the metal ion from the phase in which it occurs, hence cannot involve processes such as precipitation, solvent extraction, or the use of ion-exchange resins as means to this end. Sequestration is thus almost synonymous with masking.

Because masking is often achieved by pH adjustment or by adding excess of a suitable complexing agent to the solution, the procedure may appear to be deceptively simple. The qualitative factors involved in the selection of masking agents can be readily understood from the properties of metal ions and complexing species[7]; nevertheless, the quantitative treatment of equilibria in such systems has, until recently, been considered too difficult so that masking procedures have been very largely empirical. However, Ringbom[8] has developed a mathematical approach to metal-complex equilibria which can be applied to masking and demasking. Ringbom's approach is a general one and requires little modification in passing from one type of analysis to another. This is because the underlying principles involved in masking change only marginally with the methods used. Thus in spectrophotometry there is the additional requirement that the masked and the analytically important species must not absorb light in the same region of the spectrum, whereas in gravimetric analysis it is necessary that the masked species is not coprecipitated. Similarly, in ion exchange masked and not-masked species must behave differently toward an ion-exchange resin, and in oxidation-reduction titrations they must have different oxidation-reduction potentials.

The advent of electronic digital computers and the availability of the necessary stability constant data have also made possible a more precise definition of optimum working conditions for many chemical systems. These aspects are discussed in Chapter 2.

REFERENCES

1. Definitions given in the Oxford English Dictionary include "to disguise; to conceal from view; to hinder from action; to conceal the real nature, intent, or meaning of"; *cf.* Shakespeare in *Macbeth,* "masking the business from the common eye," and in *Love's Labour Lost,* "most immaculate thoughts, Master, are mask'd under such colours."
2. F. Feigl, *Specific and Special Reactions,* Elsevier, New York, 1940.
3. F. Feigl, *Ind. Eng. Chem., Anal. Ed.* **8,** 401 (1936).
4. *Der Grosse Brockhaus,* 16th ed., F. A. Brockhaus, Wiesbaden, 1955.
5. Useful tabulations of methods for protecting groups in organic molecules during syntheses or degradation are given in the following references: H. J. E. Loewenthal, *Tetrahedron,* **6,** 269 (1959); R. A. Boissonas, *Advan. Org. Chem.* **3,** 159 (1963); J. F. W. McOmie, *Advan. Org. Chem.* **3,** 191 (1963); G. A. Swan, in K. W. Bentley, ed., *Elucidation of Structures by Physical and Chemical Methods,* Interscience, New York, 1963, Part 1, Chap. 8.
6. R. E. Hall, U.S. Pat. 1,956,515 (1934); Re-issue 19719 (1935).
7. D. D. Perrin, *Organic Complexing Reagents,* Interscience, New York, 1964.
8. A. Ringbom, *Complexation in Analytical Chemistry,* Interscience, New York, 1963.

CHAPTER

2

FACTORS GOVERNING THE CHOICE OF MASKING AGENTS

2.1 QUALITATIVE ASPECTS

One of the most important objectives in procedures for inorganic analysis is the achievement of selectivity. There are two major ways in which this can be done. The first of these is to develop reagents that are highly selective. The other is to add masking agents which, by suppressing the effects of interfering species, will make the reaction much more selective or even, in some cases, specific under the chosen conditions. Masking techniques depend largely on complex formation, and the study of factors governing the stability of metal complexes is likely to be a rewarding one in guiding the selection of masking agents for new applications.

In any particular system many different factors are involved in determining the effectiveness with which an ion or a neutral molecule needs to be masked if it is not to cause interference; for example, relatively weak complex formation may be sufficient to prevent precipitation or to reduce an already low concentration of a foreign ion below the level at which it causes detectable interference, but if the same ion is present at very high concentrations much more effective masking may be necessary. Considerations such as these make it difficult to lay down firm rules: a reagent that is suitable under one set of conditions may be quite inadequate at a different pH or a different metal ion concentration. In general, however, stability constants of complexes must be sufficiently high, reaction must be rapid, and precipitation should be avoided where possible.

Qualitatively, the best ligands to use as masking agents would be those which form strong, colorless complexes with the ions to be masked but relatively weak complexes with the other cations that are present. Thus copper(II) is reduced to copper(I) by thiourea with the formation of complexes that are more stable than those of calcium, whereas the calcium complexes with EDTA are stronger than those of copper(I). Hence calcium is titrated preferentially with EDTA in the presence of copper ions if thiourea is added as a masking agent.

The selectivity of masking agents can often be improved by careful pH control or by using two or more different kinds simultaneously, such as triethanolamine for iron(III), tartrate for antimony and bismuth, and potassium iodide for mercury(II). It is also necessary to consider the particular situation in which a masking agent is to be used. Highly colored complexes are obviously unsuitable where spectrophotometric measurements have to be made, or where the end point of a titration has to be observed. Thus thioglycolic acid forms an intensely red complex with iron(III), hence is an unsatisfactory masking agent for use when much iron is present, unless triethanolamine or some other alternative complexing species is also added to mask iron against this reaction. Considerations other than purely chemical ones must sometimes be taken into account. Cyanide ion is highly poisonous, and acidification leads to the evolution of hydrocyanic acid. Again, although dimercaptopropanol is nontoxic it has a most objectionable smell.

Factors governing the stability of metal complexes have been reviewed elsewhere,[1] but it is convenient to recapitulate them briefly.

Only a small number of types of groups commonly take part in metal complex formation in aqueous systems, and these further subdivide into those where coordination of the metal is through the lone pairs of electrons on nitrogen, oxygen, or sulfur. The groups that involve oxygen are usually the carboxylate ion, $-COO^-$, and the enolate ion, $-O^-$, with weaker binding to ether oxygens and carbonyl oxygens (in aldehydes, amides, esters, and ketones). For nitrogen amino groups are the most important, followed by diazo and ring nitrogens. Sulfur ligands bind most strongly when thiolate ions are involved, but there is also much weaker bonding through thioether and —S—S— groups. It is on this slender foundation that the greater part of metal complex chemistry and almost all of the methods of masking and demasking are based.

There are, however, three factors that are directly responsible for the great range that can be achieved in the stabilities of metal complexes formed by ligands containing these groups.

1. The preferred stereochemical arrangements of bonds about metal ions.
2. The ability of many ligands to coordinate to a metal ion through more than one center, to give five- or six-membered chelate rings.
3. The "compatibility" of metal and ligand donor atoms, as expressed in the concept of "hard" and "soft" acids and bases.

Common stereochemistries of metal ions, listed by periodic family, are set out in Table 2.1. There is a growing body of evidence that in response to particularly favorable bonding conditions a metal ion

Table 2.1. Common Stereochemistries of Metal Ions in Their Complexes

Metal ion	Coordination no.	Stereochemistry
Cu(I)[a] Ag(I)[b] Au(I)[b]	2	Linear
Cu(I)[c] Ag(I)	4	Tetrahedral
Cu(II)[d] Ag(II) Au(III)	4	Square planar
Hg(I) Hg(II)	2	Linear
Be(II) Mg(II)(?) Zn(II) Cd(II) Hg(II)	4	Tetrahedral
Mg(II)(?) Ca(II) Sr(II) Ba(II) Cd(II) Zn(II)	6	Octahedral
B(III) Al(III) Ga(III) In(III)	4	Tetrahedral
Al(III) Sc(III) Y(III) lanthanides	6	Octahedral
Sn(IV) Pb(IV)	4	Tetrahedral
Si(IV) Ti(IV) Sn(II) Sn(IV) Pb(II) Pb(IV)	6	Octahedral
V(III) V(IV)	6	Octahedral
Cr(VI)	4	Tetrahedral
Cr(II)[e] Cr(III)	6	Octahedral
Mn(II) Mn(III)[e]	6	Octahedral
Co(II)[f]	4	Tetrahedral
Ni(II) Pd(II) Pt(II)	4	Planar
Fe(II) Fe(III) Co(II) Co(III) Ni(II) Ni(IV) Pt(IV) Ru(III) Rh(III) Os(III) Ir(III)	6	Octahedral

[a] If ligands are strongly basic, highly polarizing, or easily polarized.

[b] Preferred coordination number. Can also have a coordination number of 4 but less readily than Cu(I).

[c] If ligands accept π-electrons from the metal or if bonding is ionic.

[d] Can also form distorted octahedral structures (coordination number 6), with 4 short bonds and 2 long (weaker) bonds which, in the limit, passes into square planar.

[e] Distorted octahedral, if high-spin.

[f] When ligand fields are weak.

NOTE: In addition, tetravalent ions of Zr, Hf, Mo, W, and U, and also the actinides can have a coordination number of 8 for which various stereochemistries are possible.

might adopt a stereochemistry other than that given in the Table. Nevertheless, this compilation serves as a useful working guide to conditions to be expected in aqueous systems.

The additional stability conferred on metal complexes by chelate ring formation is well known. For $Cu(NH_3)_4^{2+}$ the overall stability constant is $10^{12.6}$, whereas for the corresponding bisethylenediamine copper(II)

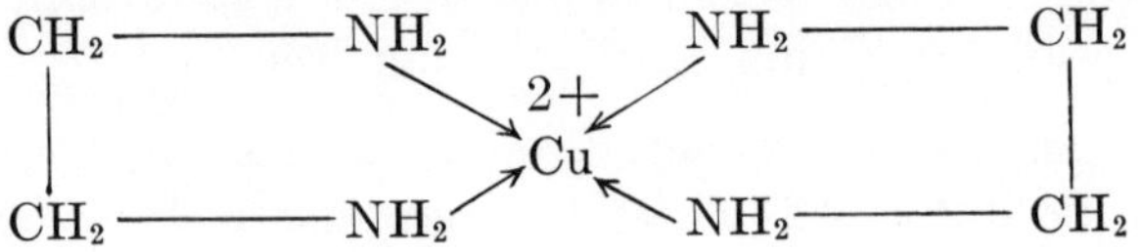

(I)

species (**I**) the constant is $10^{20.0}$, that is about ten million times greater. Cations having a coordination number of 2 can form chelate rings only with difficulty so that, in general, there is little advantage in using bidentate, rather than unidentate, ligands to mask them. Thus, although $Ag(NH_3)_2^+$ has log $\beta_2 = 7.0$, the values for the 1:1 and 2:1 ethylenediamine–silver complexes are only about 4.7 and 7.7, respectively. Contrary to the usual observation that five-membered chelate rings are more stable than corresponding but larger rings, the 1:1 complexes formed by Ag(I) with tri-, tetra-, and pentamethylenediamine are all more stable than the corresponding ethylenediamine complex, probably because the increasing ring size allows the bonds from the two amino groups to Ag to approach more closely to a linear structure.

With ter- and multidentate ligands, steric considerations are also of major importance. The more bonding sites available from the ligand, and the closer they can accommodate themselves to be preferred stereochemistry of the metal ion the stronger will be the resulting complex. The stability constants of 1:1 Cu–amine complexes illustrate this. In the series ethylenediamine, diethylenetriamine (**II**), triethylenetetramine (**III**), tetraethylenepentamine (**IV**), and pentaethylenehexamine (**V**) log K_1 values

$$NH_2CH_2CH_2NHCH_2CH_2NH_2$$

(**II**)

$$NH_2CH_2CH_2NH[CH_2CH_2NH]CH_2CH_2NH_2$$

(**III**)

$$NH_2CH_2CH_2NH[CH_2CH_2NH]_2CH_2CH_2NH_2$$

(**IV**)

$$NH_2CH_2CH_2NH[CH_2CH_2NH]_3CH_2CH_2NH_2$$

(**V**)

are 10.6, 16.0, 20.4, 22.9, and 22.4, respectively. Whereas all the amino groups of tetraethylenepentamine can be disposed about five corners of a distorted octahedron about Cu(II) ion, the sixth amino group of pentaethylenehexamine cannot, so that this group is not involved in the bonding to the metal and log K_1 does not increase.

Similarly, the copper complex with triethylenetetramine (**III**) has a higher stability constant than the copper complex with triaminotriethylamine (**VI**), whereas for the cobalt (II) complexes the reverse is true. This is in line with expectations because the amino groups of triethylenetetramine fit readily into a planar arrangement, but triaminotriethylamine is better suited to tetrahedral or octahedral complex formation.

$$\begin{array}{l} NH_2CH_2CH_2NCH_2CH_2NH_2 \\ \qquad\qquad\quad | \\ \qquad\qquad\quad CH_2CH_2NH_2 \end{array}$$

(VI)

In the extreme case of an ion like Ag^+ that usually forms linear complexes its complexes with multidentate ligands such as EDTA are relatively much less stable than the complexes of most other metal ions, which commonly have coordination numbers of four or six. This makes EDTA a very useful masking agent in analytical determinations of silver.

One of the most useful concepts for a qualitative discussion of the intrinsic factors that govern the strengths of bonds formed between metal ions and complexing species is that of "hard" and "soft" acids and bases recently developed by Pearson.[2] "Hardness" is characteristic of bonds that have a high ionic character whereas "softness" is related more to covalent bond formation.

HARD AND SOFT ACIDS AND BASES

Lewis[3] defined a base as a species which can donate a pair of electrons to form a coordinate bond, and an acid as a species which can accept a pair of electrons. In terms of these definitions all ligands are bases and all metal ions are Lewis acids.

A ligand is described as a "soft" base if its donor atom is of high polarizability, with empty, low-lying molecular orbitals. Such a donor atom is usually of low electronegativity and easily oxidized; that is, the orbitals of the valence electrons are easily distorted, often to such an extent that the electrons themselves can be removed. Conversely, in a "hard" base the donor atom is of low polarizability, hard to oxidize, has a high electronegativity, and its empty molecular orbitals are of high energy, hence are much less readily accessible.

Similarly, a metal ion is classified as a "soft" acid (in the Lewis sense) if it is of low charge, large size, and has several easily excited outer electrons. Conversely, a metal ion is a "hard" acid if it is of high positive charge, small size, and lacks easily excited outer electrons.

This approach has led to the generalization[2] that hard acids form stronger bonds with hard bases, whereas soft acids prefer to coordinate to soft bases. Such complexes are usually much more stable than related ones formed by a hard acid with a soft base, or a soft acid with a hard base.

Table 2.2. Metal Ions as Hard or Soft Acids

					Hard					
H^+										
Li^+	$Be^{\ +}$									
Na^+	Mg^{2+}	Al^{3+}	Si^{4+}							
K^+	Ca^{2+}	Sc^{3+}	Ti^{4+}	VO^{2+}	Cr^{3+}	Mn^{2+}	Fe^{3+}	Co^{3+}	Ga^{3+}	As^{3+}
Rb^+	Sr^{2+}	Y^{3+}	Zr^{4+}		MoO^{3+}				In^{3+}	
Cs^+	Ba^{2+}	La^{3+}	Hf^{4+}							
			Th^{4+}							
			U^{4+},	UO_2^{2+} and the rare earth ions						
					Intermediate					
		Fe^{2+}	Co^{2+}	Ni^{2+}	Cu^{2+}	Zn^{2+}				
		Ru^{2+}	Rh^{3+}				Sn^{2+}	Sb^{3+}		
		Os^{2+}	Ir^{3+}				Pb^{2+}	Bi^{3+}		
					Soft					
				Cu^+						
		Pd^{2+}		Ag^+	Cd^{2+}			Te^{4+}		
		Pt^{2+}, Pt^{4+}		Au^+	Hg^+, Hg^{2+}	Tl^+, Tl^{3+}				

The classification of metal ions as hard or soft acids is given in Table 2.2. This division also corresponds to classes *a* and *b* of Ahrland, Chatt, and Davies.[4] As expected, class *a* metal ions (hard acids) show the following stability sequences in their complexes:

$$N \gg P > As > Sb$$
$$O \gg S > Se > Te$$
$$F > Cl > Br > I$$

With class *b* metal ions (soft acids), the sequences are different:

$$N \ll P > As > Sb$$
$$O \ll S \sim Se \sim Te$$
$$F < Cl < Br < I$$

A corresponding grouping of ligands as hard or soft bases is made in Table 2.3. Ligands containing oxygen as the donor atom are "harder" than those containing nitrogen, and these, in turn, are much "harder" than sulfur-containing ligands.

The hardness or softness of acids and bases should not be confused with their strength or weakness as measured by the ease with which they can gain or lose protons. Fluoride ion is a weak base, and cyanide ion is a much stronger base, yet aluminum forms more stable complexes with fluoride than with cyanide. With silver(I) the converse is true. Nevertheless, a proton behaves as a very hard Lewis acid so that, provided only closely related sets of ligands are considered, a correlation

Table 2.3. Ligands as Hard or Soft Bases

				Hard					
H_2O	OH^-	RCO_2^-	PO_3^{3-}	SO_4^{2-}	CO_3^{2-}	NO_3^-	ROH	RO^-	R_2O
F^-	Cl^-								
				Intermediate					
NH_3	RNH_2	N_2H_4							
Br^-	N_3^-	NO_2^-	SO_3^{2-}	pyridine aniline					
				Soft					
R_2S	RSH	RS^-	SCN^-	$S_2O_3^{2-}$					
R_3P	$(RO)_3P$								
I^-	CN^-								

might be expected between the pK_a values of the protonated ligands and the stability constants of the corresponding complexes with hard metal ions, but not with soft metal ions. Within such sets, stability constants should increase with increasing basicity of the ligands, and this has frequently been observed.[5]

The oversimplifications inherent in this approach preclude the possibility that quantitative scales of "hardness" and "softness" of acids and bases can be devised. Even so, a number of useful generalizations emerge.

The stability of a complex of a hard metal ion with a hard base increases with the charge on the metal ion, so that $Al^{3+} > Mg^{2+} > Na^+$. With soft metal ions and soft bases the opposite is usually true: $Ag^+ > Cd^{2+} > Au^{3+} > Sn^{4+}$. Where exceptions occur, explanations are readily available. Thus Ti(III) is softer than Ti(I) because the electrons that have been removed are those which formerly screened the electrons in the d-shell. Similarly, Sn(IV) is softer than Sn(II), and As(V) is softer than As(III).

Hard metal ions coordinate best to the lightest atom of a family of elements in the periodic table. Soft acids coordinate best to one of the heavier atoms of the same family, possibly because these atoms have vacant d orbitals which may be used for the π-bonding of some of the d electrons of the metal ions that are soft acids.

These considerations lead to the expectation that cations with the electronic configurations of the inert gases, such as Na^+, Ca^{2+}, and Al^{3+}, will bind most strongly to "hard" ligands, such as anionic oxygen and fluorine, and will have much less tendency to bind to ligands such as ammonia, cyanide, or sulfide ions. Irrespective of their electronic configuration, cations carrying high charges also show similar preferences for ligands, often leading to the formation of hydroxo or oxo ions such as VO_2^+ and UO_2^{2+}.

In any related group, such as the alkali metal ions, or those of the alkaline earth metals, the stability of complexes usually decreases with increasing size of the cation, if the ligand is small, whereas large ligands bind more strongly to the larger cations. With multidentate anionic ligands, binding through oxygen, intermediate behavior may be observed: the calcium complex of EDTA or of cyclohexane-1,2-diamine-*N*,*N*,*N'*,*N'*-tetraacetic acid (CDTA) is more stable than the magnesium or the barium complexes.

Conversely, highly deformable cations such as Cu^+, Ag^+, and Au^+ form considerably stronger complexes with cyanide ion, iodide ion, and ammonia than with hydroxide ion, fluoride, and water. Zinc, cadmium, and mercury(II) ions are intermediate in their behavior with complexing species, but resemble copper(I) and silver(I) more than they do calcium, strontium, and barium.

Although exceptions occur in cases where ligand field stabilization energies or stereochemical considerations become of overriding importance, it is a useful generalization that, among ions of variable valence, lower oxidation states are favored when ligands bind through sulfur or nitrogen, whereas ligands in which binding occurs through oxygen will favor higher oxidation states. *o*-Phenanthroline complexes with ferrous ion are much more stable than with ferric ion; the converse is true for citrate ion. This ability to alter the bonding characteristics of a metal species by changing the valence of its ion finds many useful applications in the analytical chemistry of the transition metals. Thus the EDTA complexes of Mn(II) and Fe(II) have comparable stability constants, but the constant for Fe(III) is 10^{10} times greater, so that oxidation of Fe(II) to Fe(III) enables iron to be readily masked by EDTA in determinations of Mn.

The hardness or softness of a ligand can be modified by substituents attached to the donor atom. Electron-donating groups increase softness, whereas electron-withdrawing groups increase hardness.

Among the transition metal ions stabilities of their complexes usually increase in passing from the first to second or higher rows of the Periodic Table. Within the first row transition elements, the stability constants of divalent metal complexes with a given ligand almost always follow the Irving-Williams series, $Mn < Fe < Co < Ni < Cu > Zn$. In such a series as this the differences between stability constants for successive cations increase sharply with the "softness" of the ligand. This is illustrated in Figure 2.1. With a "hard" ligand such as oxalic acid, the stability constant of the 2:1 complex increases by a factor of about four hundred in going from Mn^{2+} to Cu^{2+}, whereas with the softer ligand ethylenediamine the corresponding increase is 10^{12} times greater than

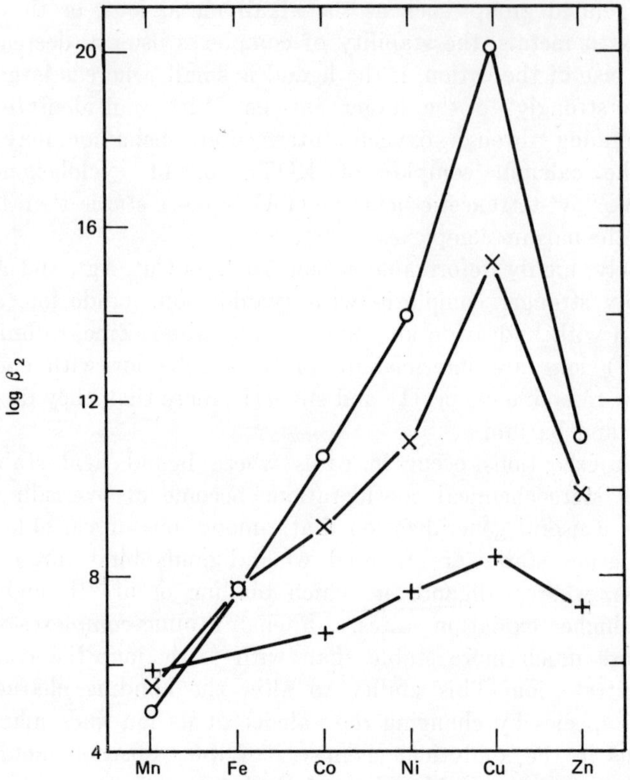

Fig. 2.1 Effect of "softness" of ligand on stability constant differences for 2:1 complexes with a series of divalent cations. Key: o, ethylenediamine; x, glycine; +, oxalic acid.

this. Intermediate values are obtained with ligands such as glycine in which bonding is through a nitrogen and an oxygen atom.

In the foregoing discussion it has been implied that all of the reactions proceed rapidly to equilibrium. However, in some cases complexes are formed or dissociate very slowly and may, in fact, persist under conditions of thermodynamic instability. Factors that confer such "inertness" on metal ions and their complexes have already been described[1] and are not dealt with further except when discussing particular masking or demasking reactions.

The fact that complexing agents are acids and bases, and hence can in many cases be used as pH buffers at the same time, is often of great advantage in analytical chemistry. This is true of simple complexing species such as acetate ion, and ammonia, as well as potential chelating agents such as ethylenediamine, glycine, tartaric acid, and citric acid. Their optimum buffer capacity is exerted near to their pK values, and this is also a region where complexing ability approaches its maximum.

2.2 QUANTITATIVE ASPECTS

THE EFFECT OF pH ON THE ABILITY OF LIGANDS TO FORM COMPLEXES

When a metal M combines reversibly with a ligand L to form a complex ML_n the equilibrium involved is expressed quantitatively by the stability constant of the complex, defined as follows:

$$\beta = [ML_n]/([M][L]^n)$$

In this expression L symbolizes the actual form that combines with the metal: for glycine this is the mono-anion $NH_2CH_2COO^-$, for ethylenediamine it is the neutral molecule $NH_2CH_2CH_2NH_2$, and for EDTA it is the tetra-anion Y^{4-}. Most complexing species however, also readily add hydrogen ions, so that at any given pH species such as HL or H_2L are also to be expected. Thus in aqueous solutions hydrogen ions and metal ions can be thought of as competing for the ligands that are present. With increase in hydrogen ion concentration (i.e., if the pH is lowered) more of the ligand molecules will be protonated so that the extent of metal complex formation will be diminished.

It is convenient to define the quantity $\alpha_{L(H)}$ to be the ratio of total protonated and nonprotonated ligand species to free ligand concentration, i.e., of $[L] + [HL] + [H_2L]$, etc., to $[L]$. The greater is this value the less efficient is the ligand as a complexing species. For ligands that can add one, two, or three protons the pH-dependence of the ratio of total ligand species to free ligand concentration is given by the following relevant expression:

$$\alpha_{L(H)} = ([L] + [HL])/[L] = 1 + 10^{(pK_1 - pH)}$$

$$\alpha_{L(H)} = ([L] + [HL] + [H_2L])/[L] = 1 + 10^{(pK_1 - pH)} + 10^{(pK_1 + pK_2 - 2pH)}$$

$$\begin{aligned} \alpha_{L(H)} &= ([L] + [HL] + [H_2L] + [H_3L])/[L] \\ &= 1 + 10^{(pK_1 - pH)} + 10^{(pK_1 + pK_2 - 2pH)} + 10^{(pK_1 + pK_2 + pK_3 - 3pH)} \\ &(= 1 + [H^+]/K_1 + [H^+]^2/K_1K_2 + [H^+]^3/K_1K_2K_3) \end{aligned}$$

where $pK_1 > pK_2 > pK_3$ are the successive pK values of the ligand.

In all expressions for β and pK, activities should ordinarily be used, but for present purposes it is satisfactory to employ "practical" constants

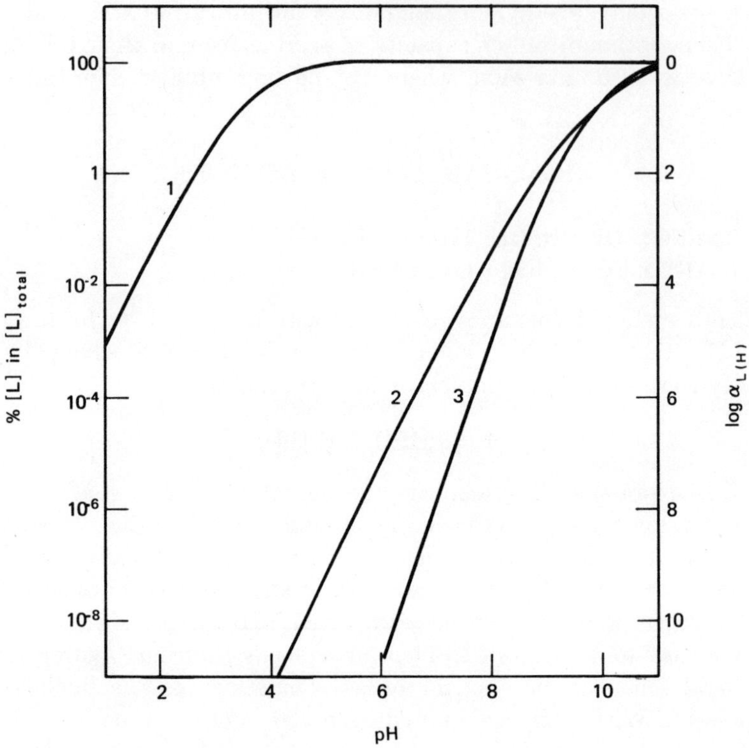

Fig. 2.2 The pH-dependence of log $\alpha_{L(H)}$ for (1) tartaric acid, (2) dimercaptopropanol, (3) triaminotriethylamine.

so long as the ionic strength at which they were obtained is not too far removed from the experimental conditions. These constants are based on concentrations of metal and ligand species, together with hydrogen ion activity as measured by a glass electrode.

When considering the effect of the pH of a solution on metal complex equilibria, it is often convenient to use, instead of the stability constant β, as defined above, "apparent" or "conditional" constants in which the free ligand concentration is replaced by the total concentration of all ligand species (including protonated forms) that are not complexed with the metal. These apparent constants are given by the following expression:

$$\beta_{\text{apparent}} = [ML_n]/([M][L]^n_{\text{total}})$$
$$= \beta/(\alpha_{L(H)})^n$$

i.e.,
$$\log \beta_{\text{apparent}} = \log \beta - n \log \alpha_{L(H)}$$

Because $\alpha_{L(H)}$ is strongly pH-dependent at pH values less than pK, an apparent constant also varies considerably with pH in the same region. In Figure 2.2 log $\alpha_{L(H)}$ is plotted against pH for three common masking agents, and indicates how, by changing the pH of a solution, the relative effectiveness of different kinds of masking agents can be varied considerably. Thus above pH 5 tartaric acid is fully ionized so that increasing the pH further will not improve its metal-binding capacity. Triaminotriethylamine, on the other hand, is virtually completely protonated at pH 5, and its effectiveness as a metal-binding species is then only about 10^{-14} of what it is at pH 11. It is clear that triaminotriethylamine will exert its greatest effect as a masking agent in alkaline solution, and demasking will usually be achieved by making the solution acid. Tartaric acid is equally effective in alkaline, neutral, or weakly acid solutions, and demasking requires much lower pH values than for triaminotriethylamine. Dimercaptopropanol is intermediate between these two in its pH dependence.

Table 2.4 lists the pK values of most of the common masking agents.

Table 2.4. pK Values of Masking Agents[a]

Acetic acid	4.7
Acetylacetone	8.9
Ammonia	9.4
Ascorbic acid	pK_1 11.3, pK_2 4.1
α,α'-Bipyridine	4.4
Boric acid	9.1
Citric acid[b]	pK_1 6.1, pK_2 4.4, pK_3 3.0
Cyclohexanediaminetetraacetic acid	pK_1 11.8, pK_2 6.2, pK_3 3.6, pK_4 2.5
1,2-Diaminopropane	pK_1 10.0, pK_2 6.9
1,3-Diaminopropane	pK_1 10.7, pK_2 9.0
Diethylenetriamine	pK_1 10.0, pK_2 9.2, pK_3 4.4
Diethylenetriaminepentaacetic acid	pK_1 10.6, pK_2 8.7, pK_3 4.4, pK_4 2.9, pK_5 1.9
Dimercaptopropanol	pK_1 10.6, pK_2 8.6
Dimercaptosuccinic acid	pK_1 10.8, pK_2 8.9, pK_3 3.5, pK_4 2.7
Ethanolamine	9.7
Ethylenediamine	pK_1 10.1, pK_2 7.3
Ethylenediaminetetraacetic acid	pK_1 10.3, pK_2 6.2, pK_3 2.8, pK_4 2.1
Ethyleneglycol bis(2-aminoethylether) tetraacetic acid	pK_1 9.5, pK_2 8.9, pK_3 2.7, pK_4 2.1
Gluconic acid	3.9
Glutamic acid	pK_1 9.2, pK_2 4.0, pK_3 2.2
Glycine	pK_1 9.7, pK_2 2.5
Hydrocyanic acid (as K salt)	9.2
Hydrofluoric acid (as Na salt)	3.1
Hydrogen peroxide	11.7

Table 2.4. (*Continued*)

Hydrogen sulfide	pK_1 14, pK_2 6.9
2-Hydroxyethylethylenediaminetri-acetic acid	pK_1 9.8, pK_2 5.4, pK_3 2.7
Hydroxylamine	6.2
8-Hydroxyquinoline-5-sulfonic acid	pK_1 8.4, pK_2 3.8
Iminodiacetic acid	pK_1 9.5, pK_2 2.7
Lactic acid	3.8
Malic acid	pK_1 4.7, pK_2 3.2
Malonic acid	pK_1 5.4, pK_2 2.7
β-Mercaptopropionic acid	4.9
Nitrilotriacetic acid	pK_1 9.8, pK_2 2.6, pK_3 2.0
Oxalic acid	pK_1 4.0, pK_2 1.1
Pentamethylenehexamine	pK_1 10.3, pK_2 9.8, pK_3 9.2, pK_4 8.6
1,10-Phenanthroline	5.0
Phenylarsonic acid	3.5
Phosphoric acid	pK_1 11.9, pK_2 6.9, pK_3 2.0
1-(2-Pyridylazo)-2-naphthol	pK_1 12.1, pK_2 1.9 (in 20% dioxan)
Pyrophosphoric acid	pK_1 8.5, pK_2 6.1, pK_3 2.5, pK_4 1.0
Salicylic acid	pK_1 13.1, pK_2 2.9
Sulfosalicylic acid	pK_1 11.6, pK_2 2.6
Tartaric acid	pK_1 4.1, pK_2 2.9
Tetraethylenepentamine	pK_1 9.5, pK_2 9.1, pK_3 8.1, pK_4 4.7, pK_5 2.7
Thenoyltrifluoroacetone	6.1
Thiocyanic acid (as K salt)	−2
Thioglycolic acid	pK_1 10.2, pK_2 3.4
Thiosulfuric acid (as Na salt)	1.4
Thiourea	2.0
Tiron[c]	pK_1 12.7, pK_2 7.7
1,2,3-Triaminopropane	pK_1 9.7, pK_2 8.0, pK_3 3.8
Triaminotriethylamine	pK_1 10.4, pK_2 9.7, pK_3 8.6
Triethanolamine	7.8
Triethylenetetramine	pK_1 10.0, pK_2 9.3, pK_3 6.8, pK_4 3.4

[a] In most cases these are "practical" constants applying at 20–25° and an ionic strength of 0.1.

[b] Also $pK \sim 16$ for the OH.

[c] Catechol-3,5-disulfonate.

From these $\alpha_{L(H)}$ values can be calculated by using the expressions given above. For the monovalent acids and bases, or for the numerically greater pK_a value of divalent acids and bases, log $\alpha_{L(H)}$ approaches zero at pH values greater than the pK_a value, whereas at pH values less than the pK_a the relationship approximates to the following:

$$\log \alpha_{L(H)} = pK_a - pH$$

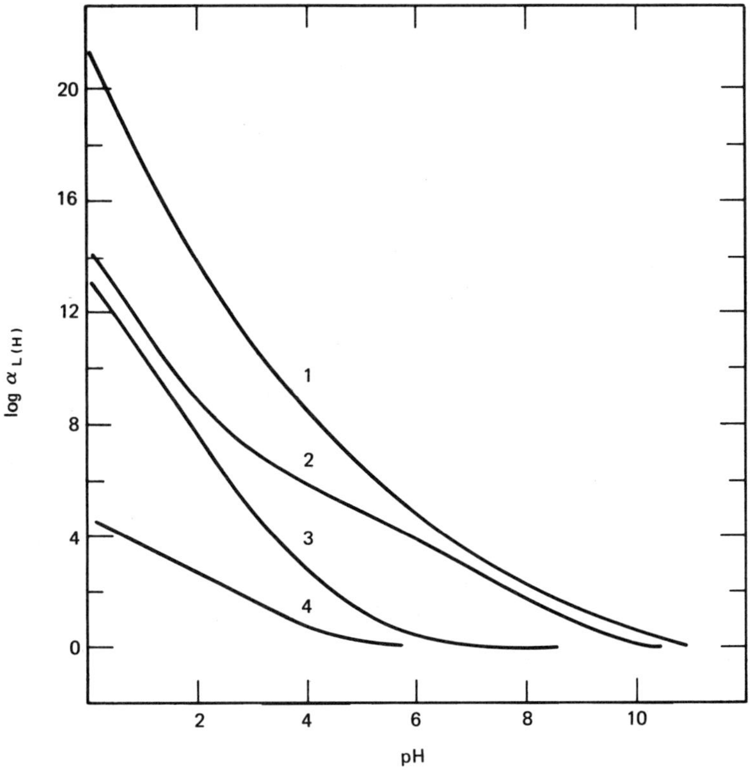

Fig. 2.3 The pH-dependence of log $\alpha_{L(H)}$ for (1) EDTA, (2) nitrilotriacetic acid, (3) citric acid, (4) acetic acid.

At pH values below the numerically smaller pK_a of divalent acids and bases the equation becomes

$$\log \alpha_{L(H)} = pK_{a1} + pK_{a2} - 2pH$$

Similar kinds of expressions can be written for the various pH regions of tri- and tetravalent acids and bases.

Figures 2.3 and 2.4 give log $\alpha_{L(H)}$ versus pH plots for some of the common masking agents.

CONDITIONAL CONSTANTS

Schwarzenbach[6] generalized the concept of conditional constants by introducing the coefficients α_M and α_L which have similar meanings to the factor $\alpha_{L(H)}$ discussed in the preceding section. The ratio of the

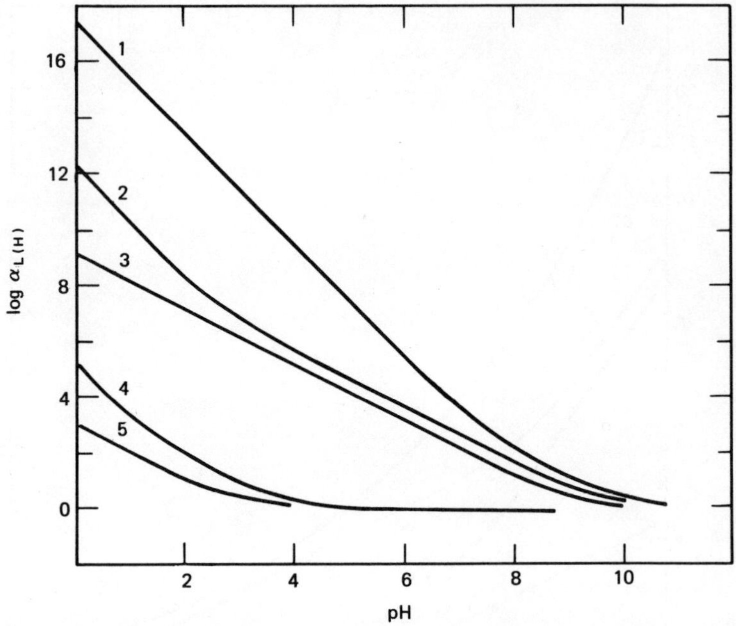

Fig. 2.4 The pH-dependence of log $\alpha_{L(H)}$ for (1) ethylenediamine, (2) glycine, (3) hydrocyanic acid, (4) oxalic acid, (5) hydrofluoric acid.

total concentration [M′] of metal ion not in the specified complex to the actual concentration of free metal ion is given by α_M, where [M′] includes free metal ion, hydrolyzed metal ion, ammono- and halogeno-metal complexes, and metal complexes with any other ligand that is present. By analogy with $\alpha_{L(H)}$ we can define $\alpha_{M(OH)}$ to measure the extent of hydrolysis of the metal ion. Similarly, [L′] includes free ligand, protonated species, and complexes with any other metals than the particular one under study.

This approach has been developed thoroughly by Ringbom,[7] who has demonstrated its usefulness for the quantitative treatment of equilibria in complexation reactions. Hulanicki,[8] and subsequently Kelly and Sutton,[9] have also discussed the quantitative treatment of masking, with particular application to precipitation reactions. In its simplest terms the problem one is commonly concerned with in analytical chemistry is to find the optimum conditions for a desired species to be determined quantitatively while, at the same time, potentially interfering species are prevented from exerting undesirable effects although they have not been physically removed from the system.

Both α_M and α_L are measures of the extent of side reactions in systems where ML_n is formed. When M reacts only with L, $\alpha_M = 1$, but if other equilibria involving M also occur, $\alpha_M > 1$. Ringbom's conditional constant is defined as follows:

$$K_{M'L_n'} = \frac{[ML_n]}{[M'][L']^n}$$

$$= \frac{\beta_{ML_n}}{\alpha_M \alpha_L{}^n}$$

If A,B,C . . . are interfering ligands

$$\alpha_M = [M']/[M] = ([M] + [MOH] + \cdots + [MA] + [MA_2] + \cdots + [MB] + \cdots)/[M]$$
$$= \alpha_{M(OH)} + \beta_{MA}[A] + \beta_{MA_2}[A]^2 + \cdots + \beta_{MB}[B] + \cdots$$

Similarly, if X and Y are other metal ions

$$\alpha_L = \alpha_{L(H)} + \beta_{XL}[X] + \beta_{YL}[Y] + \cdots$$

Evaluation of $\alpha_{M(OH)}$ and the other terms in α_M is straightforward if the necessary constants are available,[10] so long as the "interfering" ligands are present in amounts that are not significantly diminished by complex formation.

One of the difficulties in applying such quantitative treatment, however, is that the equilibrium constants, especially for the hydrolysis of most metal ions, and also for their participation in hydrolyzed-complex formation, are either unknown or not known with sufficient accuracy. In many cases the constants have been obtained at such high ionic strengths that no satisfactory corrections can be applied to convert them to more typical experimental conditions. Also, many metal ions hydrolyze to polynuclear species, the equilibrium constants of which are difficult to apply except by methods of successive approximation. The latter problem is less serious in masking reactions because, usually, free metal ion concentrations are so low that mononuclear, rather than polynuclear, hydroxo complexes tend to be favored.

Plots of log $\alpha_{M(OH)}$ versus pH for dilute solutions of many of the commoner metal ions are given in Figures 2.5 and 2.6, based on values as quoted by Ringbom.[7] Although in many cases individual constants are uncertain, the curves give at least a semiquantitative picture of hydrolytic equilibria under such conditions.

Another factor which it is difficult to allow for, in systems where two or more kinds of ligands are present, is the formation of mixed ligand–metal complexes of types such as MAB or MAB_2. These complexes are often appreciably more stable than those containing only one kind of ligand, but quantitative methods for obtaining their stability constants have only recently become available.[11]

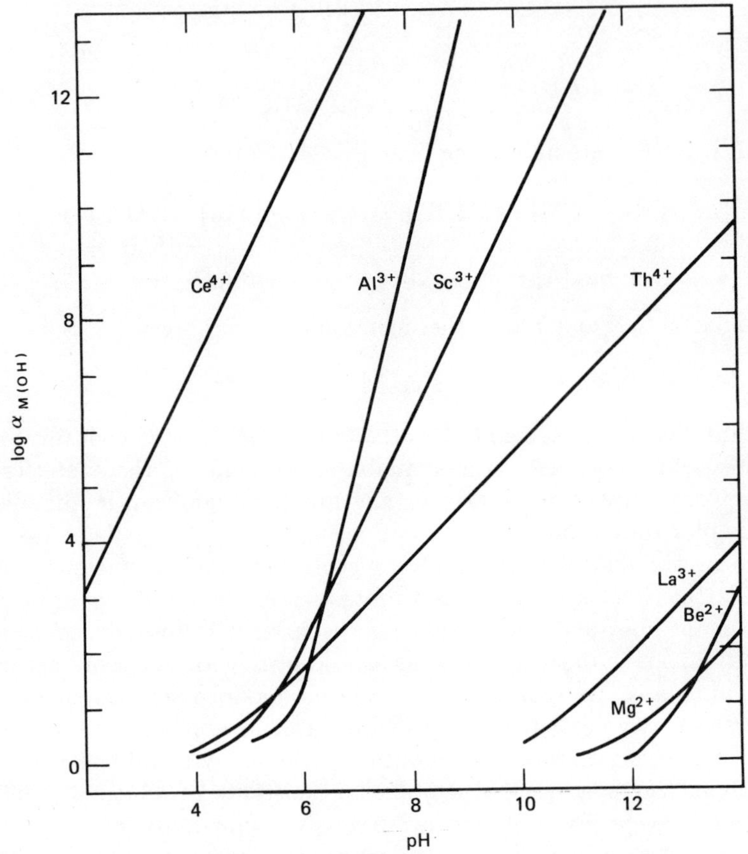

Fig. 2.5 The pH-dependence of log $\alpha_{M(OH)}$ for metal ions in groups 2A, 3A, and 4 of the Periodic Table.

In spite of these limitations this approach is a convenient one for interpreting the behavior of systems in which complex formation occurs, and its usefulness will undoubtedly increase rapidly as more equilibrium constants become available.

Where hydrolysis or competing complex formation is not important, the logarithm of a conditional constant for a complex ML has the same pH-dependence as log $\alpha_{L(H)}$. Figure 2.7, which shows the plot of log K_{BaL}' for the barium–EDTA complex is an example. Hydrolysis, however, cannot usually be neglected in considering the maximum extent to which

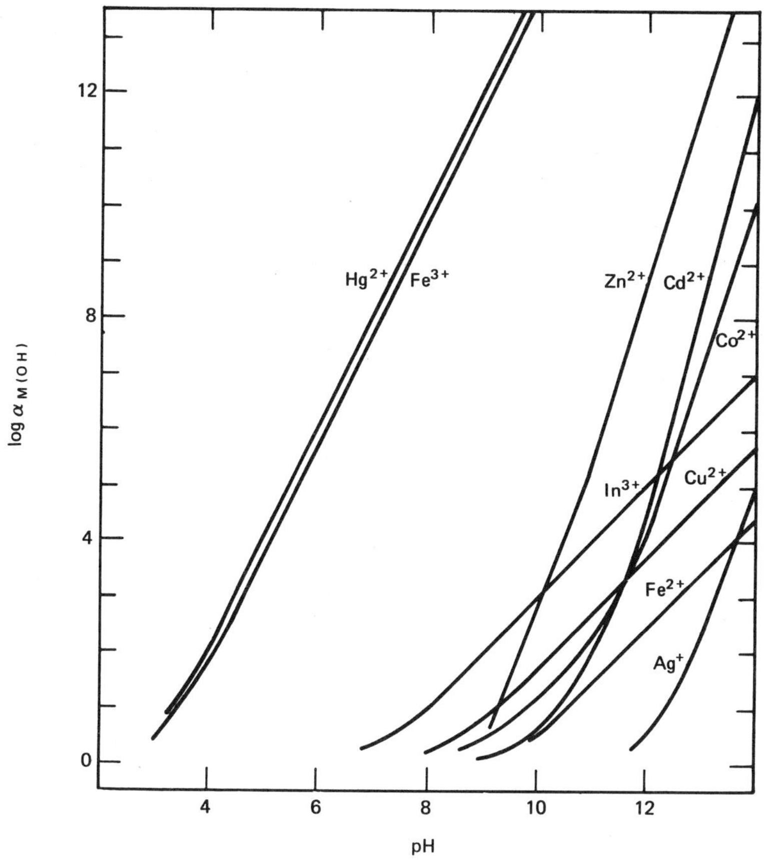

Fig. 2.6 The pH-dependence of log $\alpha_{M(OH)}$ for transition metal ions and metal ions with filled *d* shells.

a complex, ML, can be formed between M and L under conditions that the ligand can be protonated, the metal ion can be hydrolyzed, and other species, such as MHL or MOHL, can also be present. Ringbom writes the constant for such an equilibrium as $K_{M'L'(ML)'}$. The plot of log $K_{M'L'(ML)'}$ versus pH usually passes through a maximum. Results for aluminum– and cadmium–EDTA, given in Figure 2.8 are typical. Table 2.5 lists the maximum conditional stability constants of some metal–EDTA complexes, and the pH values at which they apply. These constants can readily be evaluated from the relation

$$\log K_{M'L'(ML)'} = \log \beta - \log \alpha_{M(OH)} - n \log \alpha_{L(H)}$$

which has its maximum value when $\log \alpha_{M(OH)} + n \log \alpha_{L(H)}$ is a minimum. From curves such as those drawn in Figures 2.3 to 2.6 it is thus easy to determine the optimum pH for complex formation.

The same kind of conditional constant can be applied to discuss precipitation reactions or solvent distribution equilibria if the terms involving free metal or free ligand concentrations are multiplied by the corresponding α_M or α_L values.

The problem in masking is to reduce the concentration of a specified metal ion to a sufficiently low level irrespective of what kind of, or how many, complexes are formed. For this discussion it is convenient to introduce the concept of pM, where pM, analogously to pH, is the

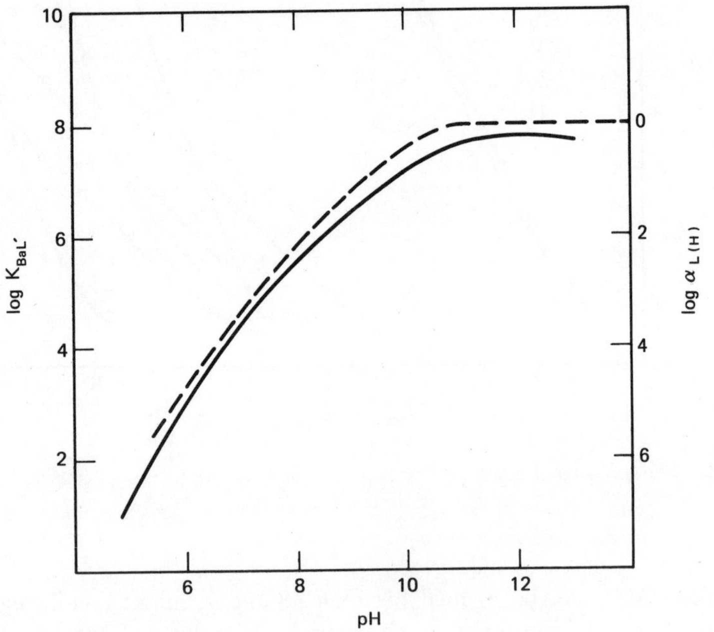

Fig. 2.7 The pH-dependence of log K_{BaL}', for the Ba–EDTA complex. For comparison, log $\alpha_{L(H)}$ is also shown (dashed line).

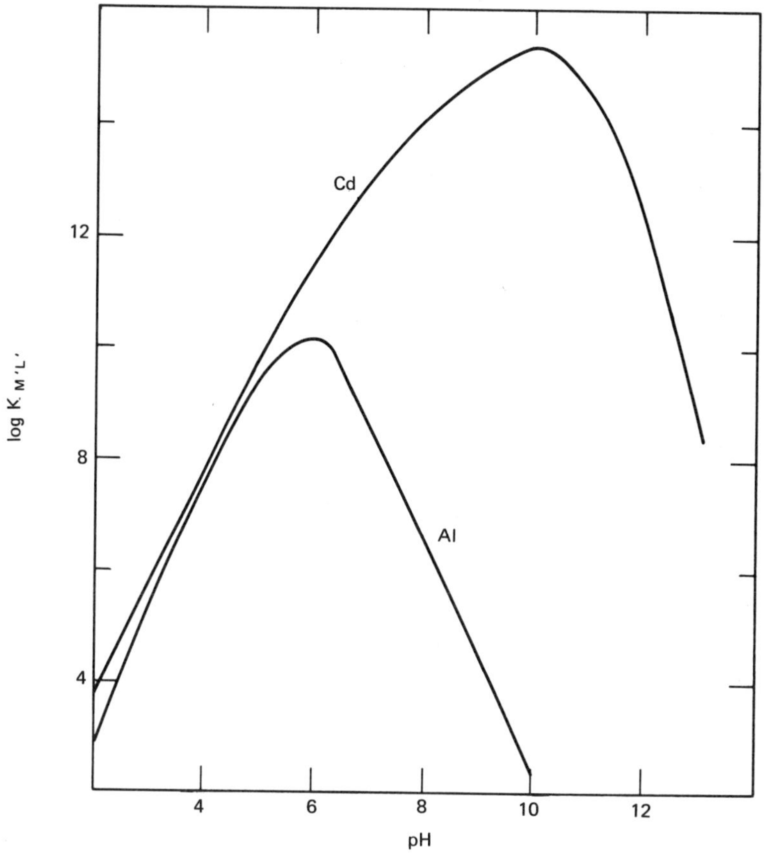

Fig. 2.8 The pH-dependence of log K_{ML}', for Al–EDTA and Cd–EDTA complexes. The curve for Al–EDTA includes acidic and basic EDTA complexes.

negative logarithm of the metal ion concentration. Thus pM = 6 corresponds to a free metal ion concentration of $10^{-6}M$ and the larger the pM value the smaller is the corresponding concentration. We define the *masking ratio,* M.R., as the ratio of the total concentration of metal species to the free metal ion concentration, and the *masking index,* M.I., as the logarithm of this ratio. Where the metal ion forms a series of complexes, such as ML, ML_2, and ML_3, and also undergoes hydrolysis, M.R. is given by the following expression:

$$\text{M.R.} = [\text{M}]_T/[\text{M}] = \alpha_{M(OH)} + \beta_1[\text{L}]_T/\alpha_{L(H)} + \beta_2[\text{L}]_T^2/(\alpha_{L(H)})^2 + \cdots$$

together with similar terms involving any other ligands that might be present. In a region in which ML_n is the predominant species this expression simplifies to

$$\text{M.I.} = \log \text{M.R.} = \log \beta_n + n \log [\text{L}]_\text{T} - n \log \alpha_{\text{L(H)}}$$

In general, three distinct conditions are to be expected. At high enough pH values, $\alpha_{\text{M(OH)}}$ will become the dominant term and only hydrolysis effects will be important. Below this region, but at pH values greater than pK_1, the masking index will be constant and equal to $\log \beta_n + n \log [\text{L}]_\text{T}$. At still lower pH values, the masking index will decrease steadily with pH by n times $\log \alpha_{\text{L(H)}}$. These effects are shown in Figure 2.9 which plots as a function of pH the masking index for an aluminum solution containing 0.1*M* fluoride. Where the p*K* of the ligand is high, and the metal ion is easily hydrolyzed, the three regions described above overlap, so that the masking index increases continuously with pH. A similar effect is observed when protonated and hydrolyzed metal com-

Table 2.5. Maximum Conditional Stability Constants for Metal–EDTA Complexes, and the pH Values at Which These Maxima Are Found

Metal ion	Log $K_{\text{M'L'}}$	pH
Ag^+	7.1	11
Al^{3+}	10.8	5.6
Ba^{2+}	7.8	12[a]
Bi^{3+}	14.1	8[b]
Ca^{2+}	10.7	12[a]
Cd^{2+}	15.6	9.7
Co^{2+}	15.0	9.6
Cu^{2+}	16.7	9.5[c]
Fe^{2+}	13.2	10
Fe^{3+}	14.8, 14.4	4–6, 11–14
Hg^{2+}	11.4	4.5[d]
La^{3+}	14.7	10.1
Mg^{2+}	8.5	11
Mn^{2+}	14.0	10.5
Ni^{2+}	17.5	9.7
Pb^{2+}	15.3	8.6
Sr^{2+}	8.6	12[a]
Th^{4+}	20.5	11–14
Zn^{2+}	15.1	8.8

[a] Almost flat from pH 11–13.
[b] Almost flat from pH 7–10.
[c] Almost flat from pH 9–10.
[d] Almost flat from pH 3–6.

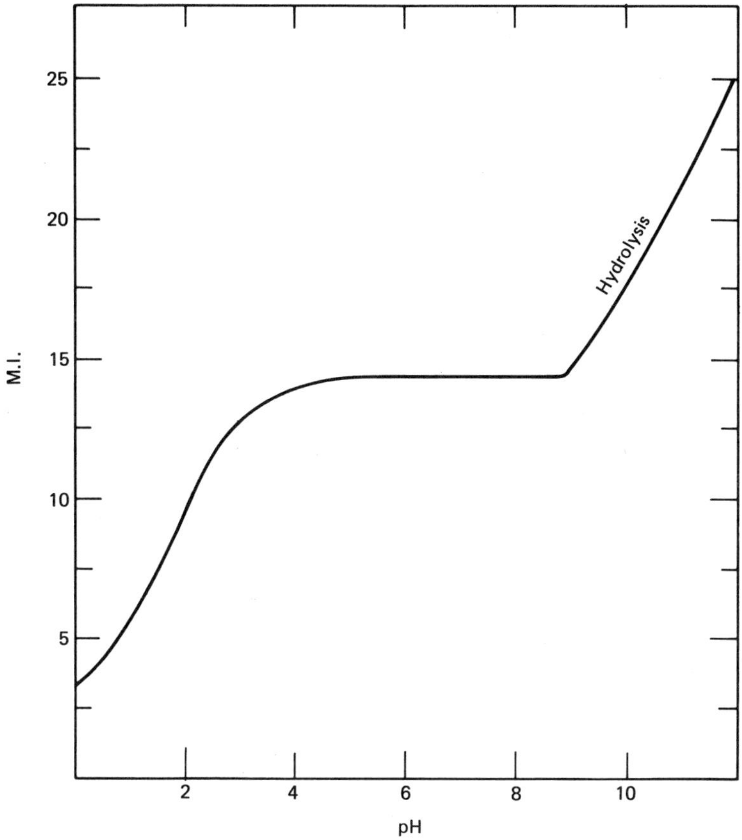

Fig. 2.9 The variation with pH of the masking index for aluminum in 0.1*M* fluoride solution.

plexes are formed, e.g., the series MH_2L, MHL, ML, M(OH)L, and $M(OH)_2L$ in the Fe(III)—EDTA system.

The treatment can be extended to include other ligands using values of $\alpha_{M(NH_3)}$ and related coefficients which Ringbom has tabulated for many metal ions and complexing species of interest to the analyst.[7]

APPLICATION OF DIGITAL COMPUTERS

The effect of a masking agent, A, on curves such as those in Figure 2.8 can readily be calculated by subtracting log $\alpha_{M(A)}$ from the values of log $K_{M'L'}$, and in this way one can assess the suitability of a masking agent for any specified application.

Recently, an alternative approach to the quantitative treatment of equilibria in metal-complex systems has become available[12,13] with the development of general computer programs for calculating the equilibrium concentrations of all species in mixtures of metal ions and complexing agents if the pH of the solution and the relevant equilibrium constants are known. Provided access can be had to a suitable computer, these programs provide a rapid method by which to ascertain the best conditions for carrying out an analytical determination involving complex formation. They are also more rigorous than is possible with manual calculations, and in one case,[13] the computer program has been adapted to deal with systems containing more than one phase, as in precipitations or solvent extractions.

The usefulness of computers for such work can be illustrated by considering the programs GENSPEC[14] (generalised species) and COMICS[12] (concentrations of metal ions and complex species). GENSPEC computes, initially for one pH value, then in specified pH increments up to a final pH value, the equilibrium concentrations of free metal ion, free ligands, and each complex species, including protonated ligands, in systems containing one kind of metal ion but more than one ligand. It prints these quantities in tabular form, then plots them as a fraction of total metal, against pH.

In this way, by taking each metal in turn one can obtain a good indication of the kinds of conditions of pH and masking-ligand concentrations in which one metal ion can best be determined in the presence of others. These conditions can be examined more closely by using the program COMICS to compute, at several pH values, all concentrations in a system which contains specified amounts of up to, say, 10 kinds of metal ions and ten kinds of ligands. One can then see, by inspection of the printed output, whether the desired reaction goes essentially to completion and whether, for example, any other metal ion or ligand would interfere under the proposed conditions. The great merit of the computer method is that no simplifying assumptions need be made so that, if the relevant constants are known, mixed-ligand, mixed-metal, polynuclear, protonated or hydrolyzed species can all be dealt with as well as the more usual ML, ML_2, ML_3 types of complexes. The following simple example is intended only as an illustration.

It required only about 1 min of computer time to assess the suitability of triaminotriethylamine as a masking agent for zinc in the complexometric titration of magnesium with EDTA. For concentrations of $10^{-3}M$ magnesium, $0.1M$ zinc, and triaminotriethylamine in $0.1M$ excess, the COMICS program showed that over the pH range 9–11 formation of MgEDTA was only 85–98.4% complete at the stoichiometric end point, hence would not provide the basis of a quantitative method. With tetraethylene pentamine, the range was 98.3–99.4% at pH 10.5. At pH 10.0 (99.3% MgEDTA), pMg was 5.2, in good agreement with the value of 5.1 calculated by Ringbom.[15] This is close to the theoretical transition point (pMg = 5.4) for eriochrome black T at this pH and supports the conclusion[15] that such a method should be satisfactory.

REFERENCES

1. D. D. Perrin, *Organic Complexing Reagents,* Interscience, New York, 1964, and references therein.
2. R. G. Pearson, *J. Am. Chem. Soc.* **85,** 3533 (1963); *Science* **151,** 172 (1966).
3. G. N. Lewis, *Valency and the Structure of Atoms and Molecules,* Chemical Catalog Co., New York, 1923, p. 141.
4. S. Ahrland, J. Chatt, and N. R. Davies, *Quart. Rev.* (*London*) **12,** 265 (1958).
5. See, for example, Ref. 1, p. 47.
6. G. Schwarzenbach, *Complexometric Titrations,* Interscience, New York, 1957.
7. A. Ringbom, *Complexation in Analytical Chemistry,* Interscience, New York, 1963.
8. A. Hulanicki, *Talanta* **9,** 549 (1962).
9. J. J. Kelly, and D. C. Sutton, *Talanta* **13,** 1573 (1966).
10. For extensive compilations of the p*K* values of aquo metal ions and the stability constants of metal complexes, see D. D. Perrin, Dissociation Constants of Inorganic Acids and Bases, *Pure Appl. Chem.* **20,** 133 (1969); L. G. Sillén, and A. E. Martell, Stability Constants of Metal-ion Complexes, *Chem. Soc.* (*London*), *Spec. Publ.* **17,** 1964; and K. B. Yatsimirskii, and V. P. Vasilev, *Instability Constants of Complex Compounds,* Pergamon Press, Inc., Oxford, 1960.
11. See, for example, D. D. Perrin, I. G. Sayce, and V. S. Sharma, *J. Chem. Soc.* (*A*) **1967,** 1755; D. D. Perrin, and V. S. Sharma, *J. Chem. Soc.* (*A*) **1968,** 446.
12. D. D. Perrin, and I. G. Sayce, *Talanta* **14,** 833 (1967).
13. N. Ingri, W. Kakolowicz, L. G. Sillén, and B. Warnqvist, *Talanta* **14,** 1261 (1967).
14. D. D. Perrin, unpublished, 1969.
15. Ref. 7, p. 148.

CHAPTER

3

MASKING AGENTS FOR CATIONS, ANIONS, AND NEUTRAL MOLECULES

The concept of *masking index,* introduced in Chapter 2, is useful for assessing the effectiveness of a masking agent in a particular situation. If log K' is the conditional stability constant for a complex formed in the principal reaction between a metal ion and a reagent, a masking agent will prevent the reaction if, under the experimental conditions, its masking index is several units greater than log K', but will not interfere if M.I. $\ll$ log K'. Masking indices are very easily calculated in regions of optimum masking, so long as stability constants are known, and correction to lower pH values usually requires only a knowledge of log $\alpha_{L(H)}$. In some cases, however, particularly when the ligand used as a masking agent is multidentate and able to form a series of metal complexes differing in their degrees of protonation, the masking index may vary with pH in a less simple manner.

Most of the masking agents in common use in analytical chemistry are not very selective in their action. They mask related cations rather than individual species. The dominant factors that determine their suitability as masking agents are (a) the kinds of atoms through which they bond to metal ions and (b) the pK_a values of the ionizable groups. Stereochemical considerations are also important, but to a lesser extent. The dependence of the stability constants of metal complexes on the "hardness" or "softness" of the bonding atoms of the ligand and the metal ion has been discussed in Chapter 2, where it was also pointed out that the maximum effect of a masking agent is exerted at pH values above the numerically greatest pK_a value of the ligand. For convenience Table 3.1 classifies masking agents into groups having similar bonding characteristics. Members within these groups are sufficiently alike that discussion of representative examples brings out their main features.

CARBOXYLIC ACIDS

From its pK_a and its position in Table 3.1 acetic acid would be expected to form its strongest complexes with "hard" cations and at pH values above 5. With most metal ions it is a rather weak monodentate

Table 3.1. Common Masking Agents, Classified by Atoms Involved in Metal Bonding

Oxygen
Hydroxide ion, hydrogen peroxide, glycols, polyols, carboxylic acids (including acetic, oxalic, malonic, malic, tartaric, citric and gluconic acids), *o*-diphenols (pyrocatechol, tiron) salicylic acid, sulfosalicylic acid, β-diketones (acetylacetone, dibenzoylmethane), sulfate, phosphates
Nitrogen
Cyanide ion, 1,10-phenanthroline, 2,2′-bipyridine, ammonia, ethylenediamine, trien, tetren
Sulfur
Sulfide ion, dithiols (dimercaptopropanol, toluene-3,4-dithiol, unithiol), potassium trithiocarbonate, sodium diethyldithiocarbamate
Oxygen and nitrogen
α-aminoacids (glycine), aminopolycarboxylic acids (NTA, EDTA, CDTA, and their derivatives), triethanolamine, *N,N*-dihydroxyethylglycine
Oxygen and sulfur
Thioglycolic acid, β-mercaptopropionic acid, dimercaptosuccinic acid
Nitrogen and sulfur
Thiosemicarbazide, cysteine, dithizone, mercaptoethylamine
Other atoms
Fluoride, chloride, bromide, iodide ions

ligand so that even in 1*M* acetate solutions masking indices for divalent cations are only about 0.5 for alkaline earths, increasing to 2–4 for transition and "palladium" (filled *d* shell) type cations such as Pb^{2+}. This is sufficient to mask appreciable amounts of lead against precipitation as lead sulfate and hence to cause interference in analytical procedures. Similarly, the use of acetate buffers could lead to appreciable errors in the complexometric titration of metal ions such as Pb(II) that do not form very stable complexes with EDTA. The acetate complexes of ter- and tetravalent cations are appreciably more stable, the corresponding values being about 8, 9, and 15 for Fe(III), In(III), and Ti(III), so that acetic acid is of limited value as a masking agent, mainly to prevent hydrolytic precipitation. It does, for example, mask Cr(III) and (when present in large excess) Th, against EDTA.

With citrate ion, the additional effect of chelation becomes important, so that between pH 6 and 10 masking indices for a 1*M* citrate solution range from 2.4 to 4.3 for the series Ba to Be. This makes citrate useful as a masking agent for preventing precipitation of calcium carbonate, calcium sulfate, or magnesium fluoride. Values for many other divalent cations are somewhat greater than this, but they are still relatively very much less than for complexes in which the ligands bond through nitrogen or sulfur. Especially with trivalent cations, further complex formation becomes possible in strongly alkaline solution because of ionization of the alcoholic-OH group of citric acid ($pK_a \sim 16$). Consequently, the masking index increases with pH (until, finally, hydrolysis supervenes), reaching values with $0.5M$ citrate solution at pH 13 of 26 for Al, 22 for Fe(III), and ranging from 8 for Zn and Cd to 15 for Cu(II). Citrate complexes are appreciably more stable than the corresponding tartrate complexes, except in strong alkali.

The insolubility of many oxalates, including those of the alkaline earths, restricts the use of oxalic acid as a masking agent. Nevertheless, in weakly alkaline solutions it is a stronger masking agent than citric acid for Al, Fe(III), Mn(III), Th(IV), and VO^{2+}. Oxalate, by acting as a demasking agent for Sn(IV) from its EDTA complex, makes possible a very selective EDTA titration procedure for Sn(IV) at pH 2.5–3.5. It is probable that rare earths, Nb, and Ta are also masked satisfactorily by oxalate.

Other carboxylic acids tend to be intermediate in character. Thus tartaric acid usually forms weaker complexes than citric acid, and complexes involving ionization of the alcoholic —OH groups have not been reported. At pH 12, tartrate retains Mg in solution during the EDTA titration of Ca. Similarly, at a lower pH it has been used to mask W, Nb, and Ta (against precipitation) during the EDTA titration of Co(II). Manganese can be titrated with EDTA in the presence of Al(III) and Fe(III) if the latter are masked by tartrate ion.

Ions such as tartrate and citrate find their main application in preventing the precipitation, in alkaline solution, of the hydroxides of polyvalent metals such as U(VI), Al, Cr, Fe, Ti, Sb, and Zr. When Mn(II) is masked in this way in ammoniacal solution, a reducing agent such as ascorbic acid or hydroxylamine hydrochloride must also be present if aerial oxidation to Mn(III), with subsequent interference, is to be avoided. These carboxylate ions are also effective masking agents when used in complexometric titrations; for example, at pH 6.5–7.5 citrate ion masks Bi, Cr, Fe(III), Mo(VI), Nb, Sb, Sn, Ta, Th, Ti, U(VI), W, and Zr, while permitting the titration of Cd, Cu, Hg, Pb, and Zn with EDTA.[1]

ACETYLACETONE

The maximum masking effect of acetylacetone is exerted above pH 9, but the acetylacetone anion is unstable in alkaline solution so that this reagent is usually employed in acid or near-neutral solution. At pH 7 masking indices for an 0.1*M* acetylacetone solution are Al(III), 13; Fe(III), 17; Ga(III), 14; Hf(IV), 17; Pu(IV), 23; Th(IV), 14; U(IV), 18; Zr(IV), 19; Be(II), 8; and U(VI), 8. Under the same conditions, for lower-valent or less strongly complexing cations, values are Cd, 0.4; Ce(III), 3; Co(II), 3; La, 2; Mn(II), 1; Ni, 4; Pb, 1; and Zn, 3.

From these figures it is easy to see why, at pH 5–6, acetylacetone masks Al, Be, Fe(III), Pd, and U(VI) while at the same time permitting the EDTA titration of Bi, Cd, Ce, Co, La, Mn, Ni, Pb, Sn, or Zn.[2] (The Fe and U complexes are colored and interfere unless the solutions are dilute.) Similarly, at pH 6.5–7.5 acetylacetone masks Al and U(VI) in the titration of rare earth cations.[1] In strong acid Mo is masked, whereas Bi can be titrated.[2]

CYANIDE ION

Although cyanide ion bonds to metals through nitrogen, its behavior places it among the "soft" ligands showing very little tendency to bond to alkaline earths, rare earths, or polyvalent cations. Its most stable complexes are with lower valent, "soft" cations comprising Groups IB, IIB, and VIII of the Periodic Table, and for an 0.1*M* potassium cyanide solution at pH 9, masking indices lie in the sequence

Hg(II), 38 > Pt(II), 37 > Au(I), 36 > Tl(III), 31 > Ni(II), 27 > Cu(I), 26 > Ag(I), 19 > Cd(II), 15 > Zn(II), 13 > Pb(II), 6

Many of these numbers are subject to large experimental uncertainty but are at least of the right order of magnitude. In addition, the adoption of low-spin states by the cyano complexes $Co(CN)_6^{3-}$ and $Fe(CN)_6^{4-}$ contributes appreciably to their stabilization.

Hence cyanide ion added in excess masks Ag, Au, Cd, Co, Cu, Fe, Hg, Ni, Pd, Pt, Tl, and Zn against titration in alkaline solution with EDTA, but does not affect the titration of Al, Bi, Mg, Mn, Pb, the alkaline earths, or the rare earths. Copper(II) is reduced by cyanide ion to Cu(I) or, better, the evolution of cyanogen can be avoided by prior reduction of copper(II) with a reducing agent.

To avoid hazards due to the formation of hydrocyanic acid (pK_a 8.8), cyanide ion is used only in alkaline or weakly acidic solutions. Because of the weakness of hydrocyanic acid, as an acid, solutions of alkali cyanides are quite alkaline so that, in using them as masking agents, it is generally necessary to include a buffer to ensure pH control. The usual procedure is to add tartrate or a similar weak complexing agent to the acidic solution containing the metal ions so that, when the solution is made alkaline the metals are not precipitated as their hydroxides. The solution is next buffered with ammonia and a slight excess of cyanide ion is added. If Fe(III) is present, the solution becomes very dark because of the formation of mixed ferri–ferrocyanides, but this changes to a faint straw-yellow color if the ferricyanide is reduced by heating with ascorbic acid.[2a] The resulting ferrocyanide is less reactive than ferricyanide ion which, for example, irreversibly oxidizes some azo dyes, including eriochrome black T, in alkaline solution. The Co and Ni complexes are also colored, so that it is necessary to work with dilute solutions. The highly colored Co, Ni, and Fe cyanide complexes can cause difficulties in the visual detection of end points. By careful addition of formaldehyde (to form the cyanhydrin, $HOCH_2CN$) or chloral hydrate, Cd and Zn can be selectively demasked from their cyano complexes.

The masking action of cyanide ion can be readily overcome by acidifying the solution, and the hydrocyanic acid can be removed by boiling, **but because of the highly toxic nature of the gas, this procedure should not be undertaken except when stringent safety precautions are observed. It should be emphasized that potassium cyanide is a very toxic substance which must be handled with great care.**

AMINES

Like acetic acid, ammonia suffers from the disadvantage that it is not a very powerful monodentate ligand, so that rather high concentrations are needed for effective masking. Also, because of its high pK_a, maximum masking is observed only above a pH of 9.5, that is, in a region in which metal ion hydrolysis often has to be considered. There is little tendency to complex formation with the alkaline earths, the rare earths, or the higher valent "inert gas" types of cations, but stable complexes are formed with transition and filled *d*-shell cations. Above pH 10, in an $0.1M$ ammonia solution, some representative masking indices are Au(III), 26; Au(I), 25; Hg(II), 16; Tl(III), 13; Cu(II), 9; Cu(I), 9; Ag(I), 5; Zn(II), 5; Ni(II), 4; and Cd(II), 3. Here, again, ligand field stabilization greatly favors the formation of cobalt(III)—ammines. By comparison, constants for Fe(II), Mg(II), and Mn(II) are negligible.

With aliphatic diamines such as ethylenediamine or 1,2-diaminopropane the chelate effect leads to increased stability constants for the same kinds of metal ions, so that such amines are still effective as masking agents in neutral or weakly acid solutions; 1,10-phenanthroline and 2,2′-bipyridine can be used at even lower pH values because their pK_a

values are much lower (5 and 4.4, respectively). At pH 5.5, 1,10-phenanthroline masks Cd, Co, Cu, and Ni in EDTA titrations and can be used to improve the selectivity of titrimetric methods for Al, other trivalent cations, the rare earths, and Pb.[3] It also masks Mn(II), forming a colorless complex. For many purposes, 1,10-phenanthroline provides an alternative masking agent to potassium cyanide and is suitable for use in the pH range of 3–7. By a kind of competitive masking 1,10-phenanthroline, added in small amounts, prevents the "blocking" of xylenol orange when it is used as an indicator in the EDTA titration of Co, Cu, or of Cu and Pb.[4]

The steric effects when polyamines are used as masking agents, and the resulting differences in stability constants, have been discussed in Chapter 2. Thus triethylenetetramine selectively masks Cu(II) and Hg(II) in the EDTA titration of Pb or Zn at pH 5, using xylenol orange.[4a] At pH 10 tetraethylenepentamine masks Cu(II), Hg(II), Cd, Co(II), Ni, and Zn in the EDTA titration of Mg against solochrome black T.[4b] It has also been used as a masking agent in the EDTA titration of Ba and Pb at pH 12, against methylthymol blue.[4a]

PHENOLS

Al(III) and Ti(IV) can be masked with tiron (1,2-dihydroxybenzene-3,5-disulfonic acid) if Fe(III) is reduced to Fe(II) and complexed with cyanide ion.[2a] This is a general property of *vic*-dihydroxy compounds. Sulfosalicylic acid is also a suitable masking agent for Al(III) and U(VI), and has been used in the EDTA titrations of Fe, the lanthanides, Mn, and Zn.

THIOLS

The carboxyl groups of thioglycolic, mercaptopropionic, and dimercaptosuccinic acids have only marginal effects on their properties as masking agents, so that it is convenient to consider them with other ligands where bonding is exclusively through the sulfur atoms of thiol groups.

Many thiols have been proposed as masking agents for cations with filled or partly filled subshells. In many cases they provide suitable alternatives to highly toxic cyanide ion. Thioglycolic acid, in ammoniacal buffer, forms colorless or weakly colored complexes with Ag, Bi, Cd, Hg, In, Pb, Sn, Tl, and Zn, masking them against titration with EDTA.[5] The intense red Fe(III) complex is not formed if triethanolamine is present and sodium hydroxide is added. Cobalt and nickel also form strongly colored complexes with thioglycolic acid, but in complexometric titrations this interference can be overcome by back-titration because NiEDTA and CoEDTA do not react with thioglycolic acid.[5]

2,3-Dimercaptopropanol (BAL, dimercaprol) masks Ag, As, Bi, Cd, Hg, Pb, Sb, Sn, and Zn in the EDTA titrations of Ca, Mg and Mn(II) in ammoniacal solutions, but is less stable and more expensive than thioglycolic acid. It forms colored complexes with Co, Cu, Fe(III), Mn, and Ni, so that other masking agents must be included if these cations are present.[6] BAL will displace Co and Cu, but not Ni, from their EDTA complexes. In the presence of air, Mn(II) is oxidized to Mn(III), which forms a deep green BAL complex in ammoniacal solution. However, in the presence of a reducing agent such as hydroxylamine, this reaction does not occur and Mn(II) is not masked by BAL. In acid solution (pH 3), BAL masks Bi and Pb in the EDTA titration of Th against xylenol orange. Unithiol (2,3-dimercaptopropanesulfonic acid) is an improvement on BAL because it is more stable and has better solubility in water. It has been proposed for use in ammoniacal solution to mask As, Bi, Cd, Ge, Hg, Pb, Sb, Sn, and Zn in the EDTA titration of alkaline earths and rare earths.[7,8] Unlike BAL, unithiol is not a masking agent for Ni, so that Ni and Mn can also be titrated at pH 10 with EDTA. Demasking of Zn, and presumably other, metal complexes of unithiol can be achieved using hydrogen peroxide.

2,3-Dimercaptosuccinic acid is similar in masking behavior to BAL and unithiol, so that, at pH 10, alkaline earths can be titrated with EDTA while Cd, Co, Cu, Ni, and Pb are masked. At pH 6, Zn can be selectively titrated in the presence of a limited amount of Cd.[9] Use of 3-mercaptopropionic acid makes it possible to titrate Ca, Mg, Mn, and Ni in the presence of Bi, Co, Cu, Fe and Hg,[10] or to titrate Zn in the presence of Pb.[11] 3-Mercaptopropionic acid and thioglycolic acid are useful masking agents for Pb. In the EDTA titration of Th(IV) in acid solution, mercaptosuccinic acid has been used to mask Bi and Fe(III).

Diethyldithiocarbamate ion has some disadvantages as a masking agent: it is not very stable, especially in acid solution, and its complexes are often insoluble. Other types of di-substituted dithiocarbamates that form water-soluble complexes have been suggested as masking agents for Ag, As, Au, Bi, Cd, Co, Cu, Fe, Hg, In, Mn, Ni, Pb, Pd, Sb, Sn, Te, Tl, V, and Zn.[12] Dithiocarbaminoacetic acid (prepared from glycine, carbon bisulfide, and ammonia) has been proposed as a masking agent for use in the pH region 2–6.[13] This is lower than for most sulfur-type ligands. At pH 2–3, Bi, In, and Tl are masked against EDTA titration, but not Al, Ga, or Th, whereas at pH 5–6 Cd, Hg, and Pb are masked but not La, Mn, Zn, or the rare earths,[13] making it possible to analyze mixtures such as Zn–Cd, Ni–Cd, In–Ga, In–Th, Mn–Pb, and Zn–Hg. In all cases the complexes are stable and water soluble.

Other masking agents in which bonding is through sulfur include thiourea, thiosemicarbazide, and thiocarbohydrazide. Thiourea masks Cu(II), Pt, Hg(II), and Tl(III), permitting the EDTA titration of Fe(III) (in the presence of fluoride ion), Ni, Zn, Sn(IV), and Pb. Thiocarbohydrazide also masks Cu(II) in the presence of Sn(IV), so that the latter can be determined with EDTA by back-titration at pH 2. Thiosemicarbazide has been used to mask Hg(II) during EDTA titrations of Zn, Cd, or Pb at pH 5–6, or Bi at pH 1–2, against xylenol orange. Alternatively, thiosemicarbazide can be used in a selective method for Hg(II) by displacing Hg(II) from its Hg–EDTA complex, followed by titration of the liberated EDTA with standard Zn or Pb.

TRIETHANOLAMINE

In alkaline solutions, triethanolamine can be used as a masking agent against precipitation for Fe, Al, and small amounts of Mn in EDTA titrations.[14] In most of these applications it is desirable to add the masking agent to an acid solution, then raise the pH. This avoids the formation of precipitates which often redissolve only very slowly. When a strongly ammoniacal solution of triethanolamine is used to mask Fe(III), the solution is usually intensely colored (yellow to brown) because of colloidal ferric hydroxide. However, the solution is colorless if about one-third or more equivalents of EDTA and the triethanolamine are added to a weakly acid solution, followed by strong sodium hydroxide. The EDTA is available to complex other cations in the solution, and these can be determined by back-titration, for example, with Ca using thymolphthalein as indicator.[15] Hydroxylamine hydrochloride should be added to solutions containing Mn to prevent oxidation of Mn(II) to Mn(III), which forms a strongly colored complex with triethanolamine. Alternatively, if the Mn–triethanolamine complex is prepared at pH 13 and then cyanide ion is added and the solution brought to pH 10–11 by adding dilute acetic acid, yellow $Mn(CN)_6^{3-}$ is formed. This is a suitable method for masking Mn so long as iron is not present.

Triethanolamine has been proposed for masking Bi during the EDTA titration of Ca or Ca and Mg.[16] Similarly, Sn can be masked in the determination of Pb and Zn.[17] In ammoniacal solution, triethanolamine masks Pb against precipitation but not against complexometric titration. Nickel can be determined by direct EDTA titration in strongly ammoniacal solution if Co(II) is masked by oxidation with hydrogen peroxide in the presence of triethanolamine to give a red complex.[18] This is an alternative to masking with cyanide and hydrogen peroxide in ammoniacal solution. (Cobalt forms $Co(CN)_6^{3-}$, and Ni forms a cyano complex from which it is subsequently demasked by addition of silver nitrate.)

Aquochromium complexes are often rather slow to react with ligands. This can make masking difficult. Thus masking of Cr with triethanolamine, ascorbic acid, or EDTA is carried out by boiling a slightly acid solution for 10–15 min, followed by addition of ammonia. However, the Cr complex with triethanolamine is strongly colored so that only dilute solutions of Cr can be masked in this way in complexometric titrations.[19] This method is less satisfactory than reduction with ascorbic acid, which has been used when Ca, Ni, or Mn was determined with EDTA in the presence of large amounts of Cr.[19]

EDTA

The two features that make EDTA such a versatile masking agent are that it forms stable, water-soluble chelate complexes with most metal ions and that its complexing ability is strongly pH-dependent. EDTA also has advantages of being stable to heat and of being usable in the presence of other masking agents. One of the earliest examples of EDTA as a masking agent was its use in an acetate buffer to mask a wide range of cations when Mo, V, and W were precipitated with 8-hydroxyquinoline.[20] Many examples of such applications are given in Chapters 5, 6, and 7. An advantage that EDTA has, as a masking agent, is that metal ions can be selectively displaced from their EDTA complexes, for example by adding excess calcium ions. Also, unlike most metal ions, univalent cations such as Ag^+, Hg_2^{2+}, Tl^+, and also Be^{2+}, have very little tendency to form complexes with EDTA or other aminopolycarboxylic acids, so that EDTA can be used to mask almost all other metal ions in their determination. This is also true where the species to be detected or determined is an anion.

Uranyl ion forms only a weak EDTA complex, so that when U(VI) is extracted into tributyl phosphate/chloroform, EDTA is a suitable masking agent for the cations of Fe, Co, Ni, Cu, Zn, and Mn. At the same time the addition of calcium ion ensures that there is no residual EDTA.[21]

EDTA masks all ions except Cu and Bi that would otherwise form colored complexes with sodium diethyldithiocarbamate or diethylammonium dithiocarbamate, and many spectrophotometric methods using this fact have been described for these two metals.[22] It also prevents the precipitation of Bi and Pb as their iodides, Ag and Pb as their chromates from acetic acid medium, and Ca, Sr and Ba as their sulfates or oxalates from ammoniacal solution. On the other hand, by making the solution more acid, the masking ability of EDTA is diminished, so that in acetic acid solutions EDTA does not mask Ba or Tl against precipitation with chromate ion, or Ca, Sr, and Ba with sulfate or oxalate. Similarly Co, Ni, and Pb (but not Tl, Ag, Hg, Bi, As, Sb, or Sn) are masked against precipitation by sulfide ion, but they can subsequently be demasked by adding excess calcium ion.

A disadvantage of EDTA as a masking agent in spectrophotometric methods where measurements are made directly on the aqueous phase is that many metal ions form colored complexes with EDTA if they are present in appreciable amounts. These include Co(II), Cr(III), Cu(II), Dy(III), Ho(III), Mn(II), Mo(V,VI), Nd(III), Ni(II), Pd(II), Pt(II,IV), Ru(III,IV), Ti(IV)) U(IV,VI), and V(IV,V).

The metal-binding properties of EDTA are shared by many other aminopolycarboxylic acids, such as nitrilotriacetic acid, which can act as a tetradentate chelate, and cyclohexanediaminetetraacetic acid (CDTA). The high stability constants of many EDTA–metal complexes are associated with the ability of EDTA to act as a sexadentate chelate which forms octahedral complexes with appropriate metal ions. CDTA is an analog of EDTA, but with a more rigid structure, and this increases the stability constants of almost all of its metal complexes by 2–3 logarithm units. Coordination numbers greater than 6 can be satisfied by diethylenetriaminepentaacetic acid which can form octadentate complexes. This has almost no effect on constants for Ca and Mg but improves complex formation with ter- and tetravalent cations, especially among the heavier elements. Effects of structural changes on the metal-binding properties of aminopolycarboxylic acids have been extensively studied.[23]

Nevertheless, until now, little use has been made of aminopolycarboxylic acids other than EDTA as masking agents. *o*-Carboxyaniline-*N*,*N*-diacetic acid (anthranilic acid diacetic acid) has been used to mask Hg(II) and other cations in a selective extraction procedure for Ag^+, based on partition of di(*n*-butylammonium) silver salicylate into isobutylmethylketone.[23a] Dihydroxyethylglycine masks Co(II) and Fe(III), when they are present together, in the determination of Ni by dimethylglyoxime.[23b] (On its own, Co(II) does not interfere, and Fe(III) can be masked by tartrate, but in admixture they ordinarily form a precipitate along with the nickel dimethylglyoxime.) Nitrilotriacetic acid, and most other complexones, mask Ni against precipitation with dimethylglyoxime, but by adding calcium ions to the alkaline solution, Ni(II) is displaced and the reaction proceeds normally. HEDTA has been used as a masking agent to prevent the precipitation of Fe(III) and Cu(II) phosphates during the determination of Mn in plant material: EDTA was ineffective.[23c]

Future developments of masking agents by modification of aminopolycarboxylic acids may well include the development of more cagelike molecules, and molecules containing more selective groups such as thiols and amino groups, and possibly groups in which metal binding would be through phosphorus or arsenic.

HALIDES

An advantage of halide ions as masking agents is that their effectiveness is independent of the pH of the solution, except for fluoride ion in strong acid. Thiocyanate ion offers the same advantage. Fluoride ion masks polyvalent cations such as Al, Fe(III), Ge, Hf, Sb, Sn, Th, Ti, Zr, and the rare earths, as well as Ca and Mg, usually by precipitation. Thus Ca and Mg can be masked when Zn is titrated with EDTA. It is one of the very few reagents that does not also mask Hg. Addition of boric acid or Be^{2+} demasks the above cations, converting fluoride ion to BF_4^- and BeF_4^{2-}.

Chlorides and iodides are limited in their masking ability to metals that have filled or partly filled *d*-shells, especially those of large atomic number. The feebly ionized $HgCl_2$, HgI_4^{2-}, insoluble AgI, and the precipitation of Cu(I) iodide when potassium iodide is added to a copper(II) solution are representative examples. If Cu(I) is precipitated in this way and the liberated iodine is removed by adding ascorbic acid or hydroxylamine, other metal ions in the solution can be titrated without interference from Cu(I).

The demasking of Hg(II) by iodide ion from the Hg–EDTA complex, with subsequent determination of the displaced EDTA, is the basis of an almost specific complexometric method for Hg(II).[23d]

OTHER INORGANIC ANIONS

Carbonate ion can be used to mask U(VI) in the titration of Zn at pH 8–10.5 with EDTA, but because of the insolubility of the carbonates of many metals, this procedure has little practical application. Hydroxide ion masks Al, Pb, and Zn by converting them in strongly alkaline solution to aluminates, plumbates, and zincates: they can be demasked by lowering the pH. At pH 4 orthophosphate ion masks W(VI) against titration with EDTA, and at pH 8 Fe(III), Al(III), and Cr(III) can be masked by pyrophosphate in the EDTA titration of Co(II). At pH ≤ 2, Th(IV) is masked by sulfate against titration with EDTA:

the reaction is reversible and at pH 4–5 the Th is demasked. Sulfate ion can also be used to mask Ti(IV). Thiocyanate masks Hg(II) in the EDTA titration of Bi(III) at pH 1, but the Hg(II) can be demasked and titrated at pH 6 by adding excess silver nitrate. Thiosulfate is useful for masking Cu(II), Hg(II), Pt(IV), and Tl(III) against titration with EDTA, and it also overcomes the blocking effect of Ag(I) on hydroxyazo dyes.

3.1 MASKING AGENTS FOR CATIONS

From the foregoing discussion it is apparent that the selection of a masking agent for any particular application must involve the consideration of many factors. These include the species to be masked, the species which it is desired to retain in a reactive condition, their absolute and relative concentrations, the pH of the solution, other reagents that are to be used, including indicators for titrations, and the nature of the intended principal reaction. In general, when a cation is masked, the resulting complex should have little or no color and be soluble: traces of copper(II) can be masked by diethyldithiocarbamate ion, but the intense color and insolubility of the complex would cause difficulties in most analytical procedures if appreciable amounts of copper(II) were masked in this way.

It sometimes happens that separation procedures are often difficult or impracticable in cases where a component of a sample has to be determined in the presence of large amounts of other compounds. Although selectivity might be improved by the use of suitable masking agents, a large excess of a particular masking agent commonly has a detrimental effect on the principal reaction so that the gain in selectivity is at the expense of sensitivity. The realization that an optimum concentration of masking agent existed for such systems led to the introduction of the concept of *substoichiometric masking*,[24] in which the amount of masking agent added is less than the amount needed for complete masking of the interfering substance but is, instead, the amount that leads to least disturbance of the principal reaction.

The following two examples illustrate the way substoichiometric masking can be used.[24] In the polarographic analysis of Zn in the presence of up to fifty times as much Cd, addition of a ligand that preferentially complexes with Cd will diminish the height of the Cd wave on which the Zn wave is superimposed, thereby facilitating the determination. On the other hand, excess of the ligand would also complex the Zn. The same considerations apply to the determination of Co in the presence of large amounts (up to 2000 times) of Ni by adding 98–99% of the amount of EDTA needed to mask the Ni. Cobalt can then be extracted into chloroform as its green Co(III)-PAN complex, and determined spectrophotometrically.

In many cases also the only satisfactory approach, other than to have recourse to direct experiment, is to carry out the necessary quantitative calculations as outlined in Chapter 2. Nevertheless it is useful to have a list which suggests the kinds of masking agents that are ordinarily effective for individual metal ions. This is provided in Table 3.2.

Although not indicated in this Table, the properties of a masking agent can often be modified by the simultaneous addition of a metal ion that is intermediate in complex-forming ability between the cation

Table 3.2. Masking Agents for Cations[a]

Ag	CN^-, I^-, Br^-, Cl^-, SCN^-, $S_2O_3^{2-}$, NH_3, thiourea, TGA, DDC, BHEDTC, citrate
Al	F^-, BF_4^-, acetate, formate, citrate, tartrate, oxalate, malonate, gluconate, salicylate, SSA, tiron, EDTA, TEA, acetylacetone, BAL, OH^-, mannitol
As	S^{2-}, BAL, unithiol, citrate, tartrate, $NH_2OH.HCl$, OH^-
Au	CN^-, I^-, Br^-, Cl^-, SCN^-, $S_2O_3^{2-}$, NH_3, thiourea, BHEDTC
Ba	CDTA, EDTA, EGTA, citrate, tartrate, DHG, F^-, SO_4^{2-}
Be	citrate, tartrate, EDTA, tiron, SSA, acetylacetone, F^-
Bi	I^-, SCN^-, $S_2O_3^{2-}$, F^-, OH^-, DDC, TGA, unithiol, BAL, BHEDTC, MPA, DMSA, cysteine, dithizone, thiourea, citrate, tartrate, oxalate, tiron, SSA, NTA, EDTA, PDTA, TEA, DHG, triphosphate, ascorbic acid
Ca	NTA, EDTA, EGTA, DHG, tartrate, F^-, BF_4^-, polyphosphate
Cd	I^-, CN^-, $S_2O_3^{2-}$, SCN^-, DDC, BHEDTC, BAL, unithiol, cysteine, MPA, DMSA, DMPA, BCMDTC, dithizone, TGA, citrate, tartrate, malonate, glycine, DHG, NTA, EDTA, Pb-EGTA, NH_3, tetren, 1,10-phen
Ce	F^-, PO_4^{3-}, $P_2O_7^{4-}$, citrate, tartrate, DHG, NTA, EDTA, tiron, reducing agents
Co	CN^-, SCN^-, $S_2O_3^{2-}$, F^-, NO_2^-, citrate, tartrate, malonate, tiron, glycine, DHG, TEA, EDTA, TGA, DDC, BHEDTC, DMPA, DMSA, MPA, BAL, NH_3, en, tren, tetren, penten, 1,10-phen, dimethylglyoxime, H_2O_2, triphosphate
Cr	formate, acetate, citrate, tartrate, tiron, SSA, DHG, NTA, EDTA, TEA, F^-, $P_2O_7^{4-}$, triphosphate, SO_4^{2-}, NaOH + H_2O_2, oxidation to CrO_4^{2-}, reduction with ascorbic acid
Cu	NH_3, en, trien, tetren, penten, 1,10-phen, tartrate, citrate, tiron, glycine, DHG, picolinic acid, NTA, EDTA, HEDTA, S^{2-}, TGA, DDC, DMSA, DMPA, MPA, BCMDTC, BHEDTC, BAL, thiosemicarbazide, thiocarbohydrazide, cysteine, CN^-, thiourea, $S_2O_3^{2-}$, SCN^- + SO_3^{2-}, I^-, ascorbic acid + KI, $NH_2OH.HCl$, $Co(CN)_6^{3-}$, NO_2^-

Table 3.2 (*Continued*)

Fe	tartrate, oxalate, malonate, NTA, EDTA, TEA, acetylacetone, tiron, SSA, DHG, OH^-, F^-, PO_4^{3-}, $P_2O_7^{4-}$, S^{2-}, trithiocarbonate, $S_2O_3^{2-}$, BAL, DMSA, MPA, BHEDTC, TGA, oxinesulfonic acid, CN^-, reduction with ascorbic acid, $NH_2OH.HCl$, SO_3^{2-}, $SnCl_2$, sulfamic acid, thiourea, 1,10-phen, 2,2′-bipyridine
Ga	citrate, tartrate, oxalate, SSA, EDTA, OH^-, Cl^-, unithiol
Ge	oxalate, tartrate, F^-
Hf	oxalate, citrate, tartrate, NTA, EDTA, CDTA, TEA, DHG, PO_4^{3-}, $P_2O_7^{4-}$, F^-, SO_4^{2-}, H_2O_2
Hg	CN^-, Cl^-, I^-, SCN^-, $S_2O_3^{2-}$, SO_3^{2-}, tartrate, citrate, NTA, EDTA, TEA, DHG, cysteine, TGA, BAL, unithiol, thiourea, DDC, BHEDTC, MPA, DMSA, thiosemicarbazide, tren, penten, reduction with ascorbic acid
In	tartrate, EDTA, TEA, F^-, Cl^-, SCN^-, TGA, unithiol, thiourea
Ir	CN^-, SCN^-, citrate, tartrate, thiourea
La	tartrate, citrate, EDTA, tiron
Mg	citrate, tartrate, oxalate, tiron, glycol, NTA, EDTA, CDTA, TEA, DHG, OH^-, F^-, BF_4^-, $P_2O_7^{4-}$, hexametaphosphate
Mn	citrate, tartrate, oxalate, tiron, SSA, NTA, EDTA, CDTA, TEA, DHG, F^-, $P_2O_7^{4-}$, triphosphate, CN^-, BAL, oxidation to MnO_4^-, reduction to Mn(II) with $NH_2OH.HCl$ or N_2H_4
Mo	citrate, tartrate, oxalate, tiron, NTA, EDTA, CDTA, DHG, F^-, triphosphate, H_2O_2, SCN^-, mannitol, oxidation to MoO_4^{2-}, ascorbic acid, $NH_2OH.HCl$
Nb	citrate, tartrate, oxalate, tiron, F^-, OH^-, H_2O_2
Nd	EDTA
(NH_4^+)	HCHO
Ni	citrate, tartrate, malonate, NTA, EDTA, SSA, DHG, glycine, picolinic acid, F^-, CN^-, SCN^-, DDC, BCMDTC, BHEDTC, TGA, DMSA, DMPA, NH_3, tren, penten, 1,10-phen, dimethylglyoxime, triphosphate
Np	F^-
Os	CN^-, SCN^-, thiourea
Pa	H_2O_2
Pb	acetate, citrate, tartrate, tiron, NTA, EDTA, TEA, DHG, OH^-, F^-, Cl^-, I^-, SO_4^{2-}, $S_2O_3^{2-}$, TGA, BAL, unithiol, DMSA, DMPA, MPA, DDC, BCMDTC, BHEDTC, triphosphate, tetraphenylarsonium chloride
Pd	CN^-, SCN^-, I^-, NO_2^-, $S_2O_3^{2-}$, citrate, tartrate, NTA, EDTA, TEA, DHG, acetylacetone, NH_3
Pt	CN^-, SCN^-, I^-, NO_2^-, $S_2O_3^{2-}$, citrate, tartrate, NTA, EDTA, thiourea, NH_3
Pu	reduction to Pu(IV) with sulfamic acid

Table 3.2 *(Continued)*

Rare earths	citrate, tartrate, oxalate, EDTA, CDTA, F^-
Re	oxidation to perrhenate
Rh	citrate, tartrate, thiourea
Ru	CN^-, thiourea
Sb	citrate, tartrate, oxalate, TEA, F^-, Cl^-, I^-, OH^-, S^{2-}, $S_2O_3^{2-}$, BAL, unithiol
Sc	F^-, tartrate, CDTA
Se	F^-, I^-, S^{2-}, SO_3^{2-}, tartrate, citrate, reducing agents
Sn	citrate, tartrate, oxalate, EDTA, TEA, F^-, Cl^-, I^-, OH^-, PO_4^{3-}, TGA, BAL, unithiol, oxidation by bromine water
Sr	citrate, tartrate, NTA, EDTA, DHG, F^-, SO_4^{2-}
Ta	citrate, tartrate, oxalate, CDTA, F^-, OH^-, H_2O_2
Te	citrate, tartrate, F^-, I^-, S^{2-}, SO_3^{2-}, reducing agents
Th	acetate, citrate, tartrate, SSA, TEA, DHG, NTA, EDTA, CDTA, DTPA, F^-, SO_4^{2-}, 4-sulfobenzenearsonic acid
Ti	citrate, tartrate, gluconate, SSA, TEA, DHG, NTA, EDTA + H_2O_2, CDTA, tiron, mannitol, ascorbic acid, ferron, OH^-, SO_4^{2-}, H_2O_2, triphosphate
Tl	citrate, tartrate, oxalate, TEA, DHG, NTA, EDTA, TGA, Cl^-, CN^-, $NH_2OH.HCl$
U	ammonium carbonate, citrate, tartrate, oxalate, EDTA, F^-, H_2O_2, PO_4^{3-}
V	tartrate, oxalate, TEA, tiron, mannitol, EDTA, CN^-, H_2O_2, oxidation to vanadate, reduction with ascorbic acid or $NH_2OH.HCl$
W	citrate, tartrate, oxalate, tiron, mannitol, EDTA, CDTA, F^-, PO_4^{3-}, SCN^-, H_2O_2, triphosphate, oxidation to tungstate, reduction with $NH_2OH.HCl$
Y	CDTA, F^-
Zn	citrate, tartrate, glycol, glycerol, NTA, DHG, EDTA, CDTA, NH_3, tren, penten, 1,10-phen, glycine, CN^-, OH^-, SCN^-, $Fe(CN)_6^{4-}$, BAL, unithiol, dithizone, PAN, triphosphate
Zr	citrate, tartrate, oxalate, malate, salicylate, SSA, pyrogallol, tiron, TEA, DHG, NTA, EDTA, CDTA, F^-, CO_3^{2-}, SO_4^{2-} + H_2O_2, PO_4^{3-}, $P_2O_7^{4-}$, cysteine, arsenazo, quinalizarinsulfonic acid

[a] Abbreviations: BAL = 2,3-dimercaptopropanol; BCMDTC = bis(carboxymethyl)dithiocarbamate; BHEDTC = bis(2-hydroxyethyl)dithiocarbamate; CDTA = cyclohexanediaminetetraacetic acid; DDC = diethyldithiocarbamate; DHG = *N,N*-dihydroxyethylglycine; DMPA = 2,3-dimercaptopropionic acid; DMSA = dimercaptosuccinic acid; DTPA = diethylenetriaminepentaacetic acid; EGTA = ethyleneglycol-bis(2-aminoethylether)tetraacetic acid; en = ethylenediamine; HEDTA = 2-hydroxyethylethylenediaminetriacetic acid; MPA = β-mercaptopropionic acid; NTA = nitrilotriacetic acid; PDTA = propylenediaminetetraacetic acid; penten = pentaethylenehexamine; SSA = sulfosalicylic acid; TEA = triethanolamine; tetren = tetraethylenepentamine; tren = triaminotriethylamine; TGA = thioglycolic acid.

Table 3.3. Masking Agents for Anions and Neutral Molecules

Boric acid	F^-, tartrate and other hydroxy acids, glycols
Br^-	Hg(II)
Br_2	phenol, sulfosalicylic acid
BrO_3^-	reduction with N_2H_4, SO_3^{2-}, $S_2O_3^{2-}$, or AsO_2^-
Citrate	Ca^{2+}
CrO_4^{2-}, $Cr_2O_7^{2-}$	reduction with $NH_2OH.HCl$, N_2H_4, SO_3^{2-}, $S_2O_3^{2-}$, AsO_2^-, or ascorbic acid
Cl^-	Hg(II), Sb(III)
Cl_2	SO_3^{2-}
ClO_3^-	reduction with $S_2O_3^{2-}$
ClO_4^-	reduction with SO_3^{2-}, $NH_2OH.HCl$
CN^-	Hg(II), HCHO, $CH_3CH(OH)_2$, transition metal ions
EDTA	Cu(II)
F^-	H_3BO_3, Al(III), Be(II), Zr(IV), Th(IV), Ti(IV), Fe(III)
$Fe(CN)_6^{3-}$	$NH_2OH.HCl$, N_2H_4, $S_2O_3^{2-}$, AsO_2^-, ascorbic acid
Germanic acid	glycerol, mannitol, glucose, and other polyhydric alcohols
I^-	Hg(II)
I_2	$S_2O_3^{2-}$
IO_3^-	SO_3^{2-}, $S_2O_3^{2-}$, N_2H_4
IO_4^-	SO_3^{2-}, $S_2O_3^{2-}$, N_2H_4, AsO_2^-, ascorbic acid
MnO_4^-	reduction with $NH_2OH.HCl$, ascorbic acid, NaN_3, N_2H_4, SO_3^{2-}, $S_2O_3^{2-}$, AsO_2^-, or oxalic acid
MoO_4^{2-}	citrate, oxalate, F^-, H_2O_2, SCN^- + Sn(II)
NO_2^-	urea, sulfanilic acid, Co(II), sulfamic acid
Oxalate	MoO_4^{2-}, MnO_4^-
Phosphate	tartrate, Fe(III)
S	CN^-, S^{2-}, SO_3^{2-}
S^{2-}	$KMnO_4 + H_2SO_4$, S
SO_3^{2-}	Hg(II), $KMnO_4 + H_2SO_4$, HCHO
$S_2O_3^{2-}$	$MnO_4^{2-} + H_2O_2 + H_2SO_3$
SO_4^{2-}	Cr(III) + heat
SO_5^{2-}	$NH_2OH.HCl$, $S_2O_3^{2-}$, ascorbic acid
Se and its anions	S^{2-}, SO_3^{2-}, diaminobenzidine
Te	I^-
Tungstate	citrate, tartrate
Vanadate	tartrate

to be masked and the cation to be determined, with resulting improvement in the selectivity that can be obtained. Thus a large excess of Pb, Cd, or (to a lesser extent Zn) prevents the formation of some metal diethyldithiocarbamates, either by displacement (Bi, Mn, or Fe in alkaline solution) or by limiting the concentration of the complexing species. In this way Pb can be used to mask Bi, Ni, Co, and Fe when Cu is determined using sodium diethyldithiocarbamate.[25]

3.2 MASKING AGENTS FOR ANIONS AND NEUTRAL MOLECULES

Although a particular anion may be a suitable masking agent for a cation, the converse is not necessarily true. Thus silver ion is masked by excess cyanide ion, to form soluble $Ag(CN)_2^-$, but addition of excess silver ion to a cyanide solution leads to the precipitation of insoluble AgCN. The same is true of neutral ligands as masking agents. High concentrations of ammonia mask Cu(II) ion, but Cu(II) is not satisfactory as a masking agent for ammonia. There are, in fact, few cases in which metal ions can be used in this way.

Table 3.3 lists masking agents for anions and neutral molecules. It also includes several substances that do not strictly fall within the definition of masking agents because the masked species is destroyed in the reaction. This is true, for example, of the use of sulfanilic acid to remove nitrite ion by reaction to form a diazonium compound, or for the reaction of ammonium ion with formaldehyde to give hexamethylenetetramine. They are listed because of their practical utility.

REFERENCES

1. J. S. Fritz, M. J. Richard, and S. K. Karraker, *Anal. Chem.* **30,** 1347 (1958).
2. W. Z. Jablonski, and E. A. Johnson, *Analyst* **85,** 297 (1960).
2a. H. Flaschka, and R. Püschel, *Z. Anal. Chem.* **143,** 330 (1954); *Chemist-Analyst* **44,** 71 (1955).
3. R. Pribil, and F. Vydra, *Coll. Czech. Chem. Comm.* **24,** 3103 (1959).
4. R. Pribil, *Talanta* **3,** 91 (1959).
4a. C. N. Reilley, and R. W. Schmid, *Anal. Chem.* **31,** 887 (1959).
4b. C. N. Reilley, R. W. Schmid, and F. S. J. Sadek, *J. Chem. Ed.* **36,** 555, 619 (1959).
5. R. Pribil, and V. Vesely, *Talanta* **8,** 880 (1961).
6. R. Pribil, and Z. Roubal, *Coll. Czech. Chem. Comm.* **19,** 1162 (1954).
7. L. A. Vol'f, *Zavodskaya Lab.* **25,** 1438 (1959).
8. Y. V. Morachevskii, and L. A. Vol'f, *Zh. Analit, Khim.* **15,** 656 (1960).
9. T. Mekada, K. Yamaguchi, and K. Ueno, *Talanta* **11,** 1461 (1964).
10. K. Yamaguchi, and K. Ueno, *Talanta* **10,** 1195 (1963).

11. S. Hara, *Bunseki Kagaku* **10,** 633 (1961).
12. H. Bode, K. J. Tusche, and H. F. Wahrhausen, *Z. Anal. Chem.* **190,** 48 (1962).
13. O. Budevsky, E. Russeva, and B. Mesrob, *Talanta* **13,** 277 (1966); O. Budevsky, and E. Platikanova, *Talanta* **14,** 901 (1967).
14. R. Pribil, *Coll. Czech. Chem. Comm.* **19,** 58 (1954); R. Pribil, J. Körbl, B. Kysil, and J. Vobora, *Coll. Czech. Chem. Comm.* **24,** 1799 (1959).
15. R. Pribil, and M. Kopanica, *Chemist-Analyst* **48,** 66 (1959).
16. L. Barcza, *Acta Chim. Acad. Sci. Hung.* **29,** 156 (1962).
17. R. Pribil, *Chimia* (*Switz.*) **4,** 160 (1950).
18. R. Pribil, *Talanta* **12,** 925 (1965).
19. Ref. 5, p. 565.
20. R. Pribil, and V. Sedlar, *Chem. Listy* **44,** 200 (1950); R. Pribil, and M. Malat, *Coll. Czech. Chem. Comm.* **15,** 120 (1950).
21. K. Kasiura, and J. Minczewski, *Nukleonika* **11,** 399 (1966).
22. See, for example, K. L. Cheng, R. H. Bray, and S. W. Melsted, *Anal. Chem.* **27,** 24 (1955); K. L. Cheng, and R. H. Bray, *Anal. Chem.* **25,** 655 (1953).
23. See, for example, N. M. Dyatlova, and R. P. Lastovskii, *Uspekhi Khim.* **34,** 1153 (1965).
23a. D. Betteridge, and T. S. West, *Anal. Chim. Acta* **26,** 101 (1962).
23b. E. E. Byrn and J. H. Robertson, *Anal. Chim. Acta* **12,** 34 (1955).
23c. E. G. Bradfield, *Analyst* **82,** 254 (1957).
23d. K. Ueno, *Anal. Chem.* **29,** 1668 (1957).
24. H. Flaschka and J. Garrett, *Talanta* **15,** 589, 595 (1968).
25. M. Cyrankowska and J. Downarowicz, *Chem. Anal.* (*Warsaw*) **10,** 1015 (1965).

CHAPTER

4

DEMASKING

Demasking is the term used to describe the process by which a masked species regains its ability to take part in its normal reactions. In systems in which cations have been masked by complex formation with a suitable ligand this can often be achieved by adding a sufficient amount of another cation which has a greater affinity for the ligand concerned, so that the initially masked cation is once again liberated. The commonest methods of demasking can be classified under the following headings:

1. Displacement Reactions
2. Conversion of Masking Agent to Nonreacting Species
3. pH Adjustment
4. Destruction of Ligand
5. Physical Removal of Ligand
6. Change in Oxidation State of the Metal Ion

DISPLACEMENT REACTIONS

Many methods have been developed on the basis of the demasking of nickel(II) from $Ni(CN)_4^{2-}$. In ammoniacal solution, nickel is demasked by silver ion,

$$Ni(CN)_4^{2-} + 2\,Ag^+ \rightarrow 2\,Ag(CN)_2^- + Ni^{2+}$$

and the displaced nickel ions can be determined, for example, by titration with EDTA, using murexide as indicator. This procedure has been suggested for the complexometric titration of silver[1,2] and for the indirect determination of halides,[3] cyanide ion,[3] thiocyanate ion,[3] and arsenic as arsenate ion,[4] following precipitation as their silver salts. Palladium and gold also demask nickel from $Ni(CN)_4^{2-}$ and can be determined in this way.

The above reaction has been proposed as a means of determining nickel in cyanide silvering baths by adding ammoniacal silver chloride and aqueous silver nitrate to demask the nickel.[5] Similarly, the reaction provides a rapid test for silver halides if dimethylglyoxime is present in the ammoniacal solution because the liberated Ni^{2+} reacts with the dimethylglyoxime to form an insoluble red complex. The same test is given by Pd(II) and Hg(II) halides.[6] Instead of the nickelocyanide complex, ferrocyanide ion can be used in an ammoniacal solution, and the liberated ferrous ion is detected by its reaction with 2,2′-bipyridine.[7]

The reaction can be applied to the determination of nickel in the presence of cobalt by adding potassium cyanide and hydrogen peroxide to the dilute ammoniacal solution to form $Co(CN)_6^{3-}$ and $Ni(CN)_4^{2-}$. Nickel, but not cobalt, can be demasked and determined with dimethylglyoxime or titrated with EDTA, if silver nitrate is added to the solution.[8] In a related method, cobalt and nickel are determined using EDTA, by back-titration with standard calcium against calcein, and the nickel is displaced from Ni–EDTA by adding hydrogen peroxide (to oxidize Co(II) to Co(III)) and potassium cyanide (to form $Ni(CN)_4^{2-}$). The nickel content is given by titration of the liberated EDTA.[9] (Interference by iron and copper can be avoided by masking with triethanolamine and thioglycolic acid.)

Small amounts of silver (or, indirectly, chloride) ion can be determined by the ability of Ag^+ to demask copper(II) from copper diethyldithiocarbamate dissolved in chloroform or carbon tetrachloride:

$$Cu(DDC)_2 + 2\,Ag^+ \rightarrow 2\,AgDDC + Cu^{2+}$$

The liberated Cu^{2+} is then titrated complexometrically in a weakly acid solution,[10] or determined spectrophotometrically by difference. A similar procedure has been used for determining Hg(II).[11] Conversely, copper(II) has been determined polarographically after extracting as $Cu(DDC)_2$ with chloroform, back-extracting the Cu^{2+} by shaking with water containing excess Hg(II), and precipitating the remaining Hg(II) by reduction with hydrazine in dilute sulfuric acid.[12] A spectrophotometric method for traces of Cu(II) depends on the ability of Cu(II) to displace Zn(II) when the solution is shaken with zinc diethyldithiocarbamate dissolved in carbon tetrachloride.[13]

In alkaline medium cobalt(II) is masked by EDTA against reaction with sodium diethyldithiocarbamate, but the addition of a large excess of calcium ion displaces the cobalt so that the reaction proceeds. This is the basis of a selective method proposed for cobalt.[14] Likewise, lead diethyldithiocarbamate has been suggested as a specific reagent for copper(II) in a similar type of reaction.[15]

Fluoride ion can be converted, by adding excess Be or Al, to BeF_4^{2-} or AlF_6^{3-}. This property can be exploited to determine Zr or Hf by differential spectrophotometry, because fluoride ion masks Zr and Hf against reaction with xylenol orange to form a colored complex. When Al is added to remove fluoride ion, Zr and Hf are demasked and reaction takes place. Similarly, the chloroform extraction of calcium as its red complex with glyoxal bis(2-hydroxyanil) is prevented by the masking action of fluoride ion, but the calcium ion is demasked if Al(III) is added. This reaction has been suggested as a specific color test for aluminum.[16]

Masking by fluoride prevents hydrolytic precipitation of aluminum which is, however, precipitated by 8-hydroxyquinoline if Be(II) is also added to form complexes with the fluoride ion. Similar masking enables Nb to be separated from Ti, Fe, and Mn when the latter are precipitated as their hydroxides. Addition of boric acid removes fluoride ion and demasks Nb.[17,18] Detection of uranium(VI) by dibenzoylmethane can be made more selective by taking advantage of the masking of U(VI) by fluoride ion, and its demasking when Al(III) is added.

When aluminum ion is added to a mixture of Zr and Hf, masked as their oxalates at pH 2–2.5, Hf is demasked more readily than Zr, and precipitates as $HfO(Ox)_2$: this reaction can be used for the partial separation of these two metals.[18a]

Another example of the separation of closely related elements, using a displacement reaction for demasking, is the gravimetric determination of Nb and Ta in admixture. If a 5% Al salt solution is added to a mixture of Nb and Ta oxalates at pH 2 and the solution is heated slowly to 90°, Ta is slowly displaced from its oxalato complex and precipitates as tantalic acid, especially if tannic acid is also present. The stable Nb–oxalato complex remains in solution but can be quantitatively decomplexed at pH 4 by adding a further quantity of Al salt.[18b]

Cation displacement methods have also been suggested for separating Hf and Zr (as oxalato complexes by adding Fe at pH 2), rare earths (EDTA complexes, Co, Ni, Cu, at pH 6–8), Th and Sc (tartrates, by adding Fe at pH 6.5–6.8), Pb and Bi (EDTA, with Zn, Ca, or Mg, at pH 7–7.5), and Ba and Sr (as EDTA complexes with Mg or Co, at pH 7.5).[18b]

Other examples of selective precipitation, in all cases from their EDTA complexes, are Sb and Pb (by adding Cd^{2+}), and La and Pr (by adding Co^{2+}).[18a]

By equilibrating with a suspension of a cation exchanger, in the Ca^{2+} form, a solution of La–EDTA at pH 4.8, all of the La is displaced on to the cation exchanger, from which it can subsequently be eluted and precipitated.[18b] Similar methods permit the separation of Zr and Hf, and of Nb and Ta, from their oxalato complexes using a cation exchanger (Al^{3+} form), and of rare earths from their NTA and EDTA complexes, using cation exchangers (Co^{2+}, Cu^{2+}, or Zn^{2+}).[18b] The same principle is inherent in the application of solid calcium fluoride or oxalate as precipitants and collectors for rare earth ions. Thus at pH 4–4.5, quantitative collection of rare earth ions on the precipitate is possible in the presence of many other cations if the EDTA or citric acid complexes of the rare earths are dissociated by shaking with solid calcium oxalate.[18b] Other metals that can be collected on suitable precipitates after equilibration of their EDTA complexes are Ce, Pr, Pb, Y, Ba, Pb, Sb, Ag, Ni, and Co.[18b]

Examples of ligand replacement include the addition of sulfate ion to displace Th(IV) from its EDTA complex, and the precipitation of silver halides when bromide or iodide (but not chloride) ion is added to a solution of Ag(I) dissolved in aqueous ammonia to form $Ag(NH_3)_2^+$. The latter reaction is a sensitive test for these halides.

DISPLACEMENT REACTIONS IN COMPLEXOMETRIC TITRATIONS

Many complexometric methods based on this type of reaction have been described. The following examples are representative.

Bismuth can be titrated with EDTA in the presence of mercury(II) if the latter is masked by adding a slight excess of potassium thiocyanate to form soluble $Hg(SCN)_4^{2-}$. After the titration at pH 0.7–1.2, the mercury(II) is demasked by adding a slight excess of silver nitrate, the pH is adjusted to 5–6.5 with hexamethylenetetramine, and the mercury(II) is titrated with EDTA. Alternatively, if the solution contains a large excess of thiocyanate ion, the mercury(II) remains masked at pH 5–6, permitting the titration of Pb, Zn, Cd, Co, and Ni in the presence of Hg(II).[19]

Another good masking agent for Hg(II), with which it forms a 2:1 complex, is thiosemicarbazide, so that it is possible, in its presence, to mask Hg in the titration of Th, Bi, Pb, Cd, or Zn with EDTA, using xylenol orange as indicator. Alternatively, solutions containing Hg and other metals can be titrated, followed by the addition of thiosemicarbazide to displace Hg from Hg–EDTA, and back-titration of the liberated EDTA gives the Hg content. This method has been proposed for analyzing mixtures such as Hg–Cd, Hg–Pb, and Hg–Zn.[20] Potassium iodide is also a masking agent for Hg in EDTA titrations, and a method suggested for the determination of Cu and Hg uses back-titration of excess EDTA with standard copper(II), against murexide, followed by addition of potassium iodide to displace Hg from its EDTA complex. The free EDTA is then titrated with further Cu(II).[21] Thiosulfate ion can be used instead of potassium iodide.

A method for the determination of zinc in the presence of manganese and the alkaline earths is based on an initial titration with EDTA, followed by the addition of potassium cyanide and back-titration with standard Mg of the EDTA liberated from the Zn–EDTA. Alternatively, if the initial titration is carried out at pH 5–5.5, Zn can be displaced from Zn–EDTA by adding 1,10-phenanthroline, which is a stronger masking agent.[22] Cadmium is also masked under the same conditions, and this technique has been used in the EDTA titration of Bi, Pb, and Cd.[23]

Advantage can be taken of the insolubility of cadmium diethyldithiocarbamate to determine cadmium in the presence of zinc. After an EDTA titration in ammoniacal solution to obtain the total amount of cadmium and zinc, cadmium is masked by precipitation as $Cd(DDC)_2$ and the liberated EDTA is titrated.[24] An alternative masking agent for cadmium in the complexometric titration of Cd–Zn mixtures is Pb–EGTA (giving Cd–EGTA and $PbSO_4$).[25]

The same demasking methods can, in principle, be applied to systems where more than one masking agent is involved, so long as the latter are sufficiently different in nature. For example, in a Cd–Zn–Ni–Mg or Cd–Zn–Pb mixture the total metal concentration is obtained by adding excess EDTA to an ammoniacal solution and back-titrating with Mg, against methylthymol blue. Addition of diethyldithiocarbamate then masks Cd, and the liberated EDTA, titrated with Mg, gives the amount of Cd originally present. Next, thioglycolic acid masks Ni and Pb, displacing them from their EDTA complexes, and, finally, cyanide ion masks Zn.[26] The same procedure, using successive masking with thioglycolic acid and cyanide ion, enables the EDTA titration of Mn–Co–Ni or Mg–Pb–Zn–Ca mixtures.

CONVERSION OF MASKING AGENT TO NONREACTING SPECIES

In EDTA titrations, selectivity for zinc and cadmium is usually low, and only a few interfering cations can be eliminated by masking. These include Al and Ca (by boiling with ammonium fluoride), Cu (by addition of thiourea, thiosulfate, or ascorbic acid plus potassium iodide), Fe in acid media (by addition of acetylacetone or fluoride ion), and, in alkaline media, Fe and Al (by triethanolamine). However, in mixtures containing Ag, Co, Hg, and Ni, besides Zn and Cd, selectivity for Zn and Cd can be achieved by masking in an alkaline medium with potassium cyanide and then demasking the Zn and Cd with formaldehyde.[27] Berak[28] has defined the necessary conditions for demasking as a temperature below 15°, the lowest possible concentration of potassium cyanide, and the very careful control of the amount of formaldehyde. Otherwise the pH may rise so high, because of the reactions,

$$Zn(CN)_4^{2-} + 4\ HCHO + 4\ H_2O \rightarrow Zn^{2+} + 4\ HOCH_2CN + 4\ OH^-$$

$$CN^- + HCHO + H_2O \rightarrow HOCH_2CN + OH^-$$

that eriochrome black T, used as a metal indicator in the EDTA titration, may behave simply as an acid–base indicator. Chloral hydrate[29] or acetone can be used instead of formaldehyde. Thus in the determination of zinc with zincon cyanide ion was added to mask metal ions by forming cyano complexes. Addition of chloral hydrate removed excess cyanide ion and demasked the zinc without reacting with the other cyano metal complexes, thereby improving the selectivity of the reaction.[30] A typical application of this demasking reaction in complexometry is in the titration of a Mg–Zn–Cu mixture. Addition of EDTA in excess, followed by back-titration with standard Mg, gives the total amount of the three metal ions. Titration of an aliquot to which potassium cyanide had been added in excess gives Mg alone. Demasking by the addition of chloral hydrate or of formaldehyde and acetic acid then enables Zn to be determined from the amount of EDTA that is liberated. Kinnunen and Merikanto,[31] in one of the earliest applications of demasking of metals in chelatometry, determined zinc in brass by EDTA titration after masking Zn and Cu as their cyanides and then selectively demasking Zn by this reaction. A similar reaction is the basis of a spot test for formaldehyde; it depends on the demasking of nickel from $Ni(CN)_4^{2-}$ in the presence of dimethylglyoxime.[32] In the same way, demasking from its cyano complex by adding formaldehyde, followed by reaction with diphenylcarbazide, provides a spot test for Cd in the presence of Cu. If dimethyglyoxime is used instead, the test can be used for Ni.

Bisulfite ion reacts with iodine to give iodide ion, but the bisulfite is masked if excess of an aldehyde or a methyl ketone is added to a neutral solution, with the formation of bisulfite addition compounds. This can be used as a test, either for bisulfite or for aldehydes and aliphatic methyl ketones. An extension is to decolorize *p*-fuchsin (or some other triphenylmethane dye) with sulfurous acid and then use the demasking action of aldehydes (which combine with, and remove, the sulfurous acid that has reacted with the dye) as a test for the —CHO group.[33]

The addition of boric acid converts fluoride ion to the complex anion BF_4^-, and this has also been used in demasking reactions. Tin(IV) forms SnF_6^{2-} with fluoride ion, thereby masking against the precipitation of Sn(IV) as SnS_2. Subsequent addition of boric acid allows SnS_2 to be precipitated. Germanium behaves in the same way as tin. Fluoride ion masks Nb while Mo and W are extracted as their thiocyanate complexes from dilute hydrochloric acid into chloroform and acetone. The Nb is then demasked by adding boric acid, and Nb thiocyanate can be extracted as its ion pair with tetraphenylarsonium chloride.[34]

pH ADJUSTMENT

Because of the marked effect of pH on the apparent stability constants of most metal complexes, demasking can often be achieved by simple adjustment of pH. At pH 7 barium is masked by EDTA against precipitation as its sulfate, but at pH 5 it is no longer masked. Silver chloride is soluble in ammoniacal solutions, because of the formation of $Ag(NH_3)_2^+$, but it precipitates if the solution is acidified with nitric acid. The blue color of the cuprammonium ion, $Cu(NH_3)_4^{2+}$, is discharged if the solution is made acid. The masking effect of cyanide ion is overcome if the solution is acidified, so that Ag, Cu and Pd precipitate from solution and Cd, Ni, and Zn give the reactions of the free metal ions. In a complexometric method for zinc, other metals are titrated while zinc is masked with cyanide ion, then the solution is acidified with hydrochloric acid, acetate buffer is added, and the zinc is titrated with EDTA against xylenol orange.[35] Cobaltinitrite and silver thiosulfate complexes are also decomposed by acid to yield the free metal ions.

Many methods have been devised in which the metal ion is initially present as a complex. This complex is then dissociated by lowering the pH of the solution, so that the cation is liberated and reacts with an anion already present in excess in the solution. Alternatively, the demasked cation can be removed from the solution by passage through, or shaking with, a cation-exchange resin.

The nature of the reagent used to lower the pH of the solution may also be important. Dilute acetic acid is unsuitable for demasking complexes of W and Mo, or Ce and the rare earths, with EDTA because colloidal precipitates are obtained and appreciable reduction occurs.

Conditional constants for EDTA complex formation decrease markedly as the pH is lowered, so that EDTA is a much weaker ligand at pH 4 than it is at pH 10. Thus in a pyridine buffer ammonium borofluoride masks Ca and Mg (by precipitation as $Ca(BF_4)_2$ and $Mg(BF_4)_2$), whereas Mn can be titrated directly with EDTA, against methylthymol blue. The addition of concentrated ammonia demasks Ca and Mg so that they can also be titrated with EDTA.

Similarly, Mg and Ca can be determined in the presence of interfering metal ions, and phosphate ion, by titrations at three different pH values. After adding excess CDTA (cyclohexanediaminetetraacetic acid) to an acid solution the pH is adjusted to 5 and the excess CDTA is determined by titration with standard Zn solution, against xylenol orange. Similar titrations are carried out at pH 10 with standard Mg solution (eriochrome black T) and at pH above 12 with Ca (calcein II). At pH

5 all metals except Ca and Mg are complexed by CDTA. The difference between pH 10 and pH 5 gives Ca + Mg, and that between pH 12 and pH 10 gives Ca.[36] In the same way Th and the rare earth metals can be titrated consecutively with diethylenetriaminepentaacetic acid at pH 2.5–3.0 and 5.0–5.5, respectively.[37]

Hydroxyl ion is also a masking agent for some metal ions and, in such cases, making a solution less alkaline can be used as a demasking process even in the absence of added ligands. Commonly, hydroxides or hydroxy salts are precipitated, but the precipitates are often nonuniform and of poor quality for analytical purposes. In strong alkali, Mg is masked by precipitation as $Mg(OH)_2$, so that Ca can be titrated directly with EDTA. Making the solution less alkaline brings the Mg back into solution where it, in turn, can be titrated. More commonly, however, the reverse procedure is followed; for example, titration of a Cd–Zn mixture with EGTA (ethyleneglycol-bis(β-aminoethyl ether)-tetraacetic acid) gives the total metal present. Addition of sodium hydroxide displaces Zn from Zn–EGTA to form $Zn(OH)_4^{2-}$, and the liberated EGTA, titrated with Ca, using cresolphthalein complexone (metalphthalein) as indicator, gives the amount of Zn present. (In this method, moderate amounts of Pb are masked as a hydroxy complex anion, and Fe and Al are masked by triethanolamine.)

So, too, in the selective determination of Al and Ga by EDTA titration all cations are titrated by EDTA at pH 3–7, but when the pH is then raised above 10.5, Al and Ga are displaced and masked as aluminates and galliumates. The released EDTA is titrated with any suitable metal ion.[38]

Hydroxyl ion demasks Al from its oxalate and fluoride complexes, Fe(III) from its fluoride and thiocyanate complexes, and Zr from its fluoride, in all cases precipitating the metal as its hydroxide. Addition of sodium hydroxide removes Fe(III) and Bi(III) from their EDTA complexes and precipitates $Fe(OH)_3$ and $Bi(OH)_3$ but divalent metal ions continue to be masked. These precipitations (and also that of $Al(OH)_3$) can be carried out in ammoniacal EDTA solution if calcium ion is added in excess to displace these metals from their EDTA complexes.[39] Addition of Mg^{2+} is sufficient to displace Ti from its EDTA complex in ammoniacal solution, so that $Ti(OH)_4$ is precipitated. Similarly, EDTA masks Ni, Zn, and Mn against precipitation with ammonium sulfide: addition of calcium chloride displaces Mn and Zn from their EDTA complexes and they precipitate as their sulfides. When nitrilotriacetic acid is used instead of EDTA, Ni and Mn are masked but not Zn. Here, also, the metals are demasked if calcium chloride is added in excess.

DESTRUCTION OF LIGAND

In some cases, the masking agent can be destroyed by heat, chemical reaction, or both. Masking by thiocyanate ion can be abolished by oxidation of the ligand with alkaline hydrogen peroxide and heating. The excess hydrogen peroxide can be catalytically decomposed by adding a small amount of Fe(III). Aminocarboxylic acids such as EDTA are destroyed in acid solutions by permanganate or other strong oxidizing agents. Al can be liberated from its tartrate complex and precipitated as the hydroxide by destroying the ligand with hydrogen peroxide and Cu(II) ion. Fe(III), Pb, Ca, Ba, and Bi have been precipitated as their phosphates by oxidation of their EDTA complexes with hydrogen peroxide.[39a] Because there is no discrimination by hydrogen peroxide in its attack on these metal–EDTA complexes, EDTA is of no help in separating these metal ions under these conditions. A quantitative method for the separation of Ce from Pr and the other rare earths is based on demasking from their EDTA complexes, using hydrogen peroxide, followed by alkali addition to bring the pH to 7–9, whereupon Ce is precipitated as a hydroxy–peroxo complex.[18a] Praseodymium and the other rare earths remain in solution and can then be precipitated either as the hydroxide (at pH 11.5–12) or as the oxalate (pH 3.8–3.2).[18b]

In alkaline solution addition of hydrogen peroxide masks Ti(IV) against precipitation as $Ti(OH)_4$. Demasking occurs, and the hydroxide precipitates, if a reducing agent is added. Hydrogen peroxide can be destroyed by formaldehyde, hexamethylenetetramine, sulfite ion, or nitrite ion.

Triethanolamine is a masking agent for Fe(III) in alkaline solutions. In a fluorimetric titration of Ca, Mg, and Fe with EGTA, against calcein blue, small amounts of Fe(II) initially masked in this way at about pH 10.5 were demasked by destroying the ligand with formaldehyde, followed by hydrogen peroxide which oxidized all of the iron back to Fe(III).[40]

PHYSICAL REMOVAL OF LIGAND

Prolonged boiling with strong mineral acid (sulfuric or perchloric) can be used to remove stable but volatile masking agents such as hydrofluoric, hydrochloric, hydrobromic, and hydriodic acids. Commonly, however, gentler methods are adequate. Silver iodide, bromide, and chloride are selectively and sequentially precipitated, in that order, from a concentrated ammonia solution when the ammonia is allowed to escape progressively: the three halides can be separated by filtration over suitable pH ranges (pH 9.8–10.2, 8.6–8.7, <7.3, respectively).[41] Similarly, the pH of an ammoniacal solution can be lowered by heating the solution open to the atmosphere. Where ammonia is an undesirable reagent, alkylamines can sometimes be used instead.

In some cases, anionic ligands can be removed by passing the solution through an ion-exchange column. One such example is citrate ion which partially masks Mg in many determinations but which can be removed in this way from a solution adjusted to about pH 7. Another possibility, when the ligand acid is not readily soluble in water, is to remove it by solvent extraction from an acidified solution.

Advantages of this approach are the greater control that is possible over the experimental conditions. Thus the pH may be selected to obtain optimum relative stabilities of complexes or to ensure that one (or at most two or three) complex species predominate. Also ion exchange or extraction provides a simple method by which to decompose a complex or to replace one kind of ligand by another.

CHANGE IN OXIDATION STATE OF THE METAL ION

When a metal ion can exist in more than one valence state, there will usually be differences in the stability constants of the corresponding complexes with a given ligand. A change in the oxidation state of a metal ion may so decrease the stability of a complex that that metal ion may be no longer masked. Thiosulfate ion forms a very stable complex with copper(I) in slightly acid solution, and copper(I) does not react with PAN (1-(2-pyridylazo)-2-naphthol). If the solution is made alkaline, copper(I) thiosulfate is readily oxidized to copper(II) thiosulfate, where the metal is only weakly complexed, and copper(II) reacts quantitatively with PAN.

A separation procedure for Pr(III) and Nd(III) is based on the ease with which Pr(IV) complexes can be formed in alkaline EDTA or citrate solutions. These complexes are much less stable than the corresponding Nd(III) complexes so that, by demasking with fluoride or phosphate ion in the pH range 7.2–6, Pr is precipitated first, followed, at pH 4.2–3.5, by Nd.[18b] Similarly, Ce can be separated from other rare earths by oxidation to Ce(IV).

To date, however, changes of oxidation state have been used much more frequently for masking, rather than demasking, metal ions.

For some metals, notably U, Ce, Pr, Pb, Co, and some actinides, the ligands oxalate, fluoride, and carbonate ions also serve as precipitants if the metal ions are reduced. Nb(V) and Ta(V) can be separated by complexing with oxalic or malonic acid, followed by reduction with so-

Table 4.1. Further Examples of Demasking

Cation	Masked by	Demasked by	Application
Ag^+	NH_3	Br^-	Test for Br^-
Ag^+	NH_3	I^-	Test for I^-
Ba^{2+}	SO_4^{2-}(concd H_2SO_4)	H_2O	Pptn of $BaSO_4$
Cd^{2+}	CN^-	$HCHO + H^+$	Detection of Cd with diphenylcarbazide in presence of Cu
Cu^{2+}	$S_2O_3^{2-}$	OH^-	Detection of Cu with PAN
Cu^{2+}	thiourea	H_2O_2	EDTA titration of Cu
Fe^{3+}	PO_4^{3-}	OH^-	Pptn of $FePO_4$
Hf^{4+}	H_2O_2	Fe^{3+}	Release of Hf^{4+}
Hg^{2+}	CN^-	Pd^{2+}	Detection of Pd with diphenylcarbazide
Mg^{2+}	EDTA	F^-	Titration of Mg, Mn
Mo(VI)	F^-	H_3BO_3	Release of MoO_4^{2-}
Pd^{2+}	CN^-	HgO	Determination of Pd
Ti^{4+}	H_2O_2	Fe^{3+}	Release of Ti^{4+}
U(VI)	PO_4^{3-}	Al^{3+}	Detection of U with dibenzoylmethane
$Z4^{4+}$	F^-	Ca^{2+}	Detection of Ca with alizarin S
Zr^{4+}	H_2O_2	Fe^{3+}	Release of Zr^{4+}

dium amalgam. Nb, but not Ta, is reduced and forms an Nb(III)–oxalato or –malonato complex. Hydrolysis at pH 3–4 in the presence of tannic acid and a reducing atmosphere leads to precipitation of a tannin–tantalic acid compound, whereas Nb(III) persists in solution and is subsequently precipitated by adding Al salt solution or by oxidizing to Nb(V).[18b]

The methods described for separating La and Pr, Pr and Nd, and Ce, Pr and Nd, based on selective thermal decomposition and extraction[18a] lie outside the scope of the present book.

REFERENCES

1. H. Flaschka, and F. Huditz, *Z. Anal. Chem.* **137,** 104 (1952).
2. S. J. Gedansky, and L. Gordon, *Anal. Chem.* **29,** 566 (1957).
3. H. Flaschka, *Mikrochemie ver. Mikrochim. Acta* **40,** 21 (1953).
4. A. de Sousa, *Chemist-Analyst* **54,** 14 (1965).
5. M. Malat, and K. Holocck, *Chemist-Analyst* **50,** 115 (1961).
6. A. R. Ubbelohde, *Analyst* **59,** 339 (1934).
7. F. Feigl, and A. Caldas, *Anal. Chim. Acta* **3,** 526 (1955).
8. R. Pribil, and V. Vesely, *Talanta* **13,** 515 (1966).
9. *Ibid.* **8,** 880 (1961).

10. S. Komatsu, C. Kitazawa, and T. Hatanaka, *J. Chem. Soc. Japan, Pure Chem. Sect.* **85,** 435 (1964).
11. S. Komatsu, C. Kitazawa, and S. Nasu, *J. Chem. Soc. Japan, Pure Chem. Sect.* **85,** 598 (1964).
12. T. Fujinaga, M. Ishibashi, and K. Yamashita, *Bunseki Kagaku* **11,** 1122 (1962).
13. K. Ogiolda, *Chem. Anal. (Warsaw)* **10,** 611 (1965).
14. R. Pribil, and J. Jenik, *Coll. Czech. Chem. Comm.* **19,** 470 (1954).
15. I. Adamiec, *Rudy Metale Niezalazne* **5,** 409 (1960).
16. E. Jungreis, and A. Lerner, *Anal. Chim. Acta* **25,** 199 (1961).
17. B. Sarma, *Indian J. Technol.* **1,** 331 (1963).
18. A. I. Ponomarev, and Y. I. Bykovskaya, *Zh. Analit. Khim.* **21,** 1427 (1966).
18a. Sw. Pajakoff, *Mikrochim. Acta* **1966,** 751.
18b. Sw. Pajakoff, *Ost. Chem.-Zeit.* **67**(2), 42 (1966).
19. L. Barcza, and E. Körös, *Chemist-Analyst* **48,** 94 (1959).
20. J. Körbl, and R. Pribil, *Coll. Czech. Chem. Comm.* **22,** 1771 (1957).
21. K. Ueno, *Anal. Chem.* **29,** 1668 (1957).
22. R. Pribil, and J. Horacek, *Talanta* **14,** 313 (1967).
23. R. Pribil, and M. Kopanica, *Chemist-Analyst* **48,** 87 (1959).
24. R. Pribil, *Coll. Czech. Chem. Comm.* **18,** 783 (1953).
25. R. Fabregas, A. Prieto, and C. Garcia, *Chemist-Analyst* **51,** 77 (1962).
26. R. Pribil, *Komplexometrie I,* 2nd ed., Verlag für Grundstoffindustrie, Leipzig, 1963.
27. H. Flaschka, and F. Huditz, *Z. Anal. Chem.* **137,** 172 (1952); H. Flaschka, *Z. Anal. Chem.* **138,** 332 (1953).
28. L. Berak, *Hutnicke Listy* **12,** 817 (1957).
29. R. Pribil, *Chem. Listy* **47,** 1173 (1953); also ref. 24.
30. J. A. Platt, and V. M. Marcy, *Anal. Chem.* **31,** 1226 (1959).
31. J. Kinnunen, and B. Merikanto, *Chemist-Analyst* **41,** 76 (1952).
32. P. W. West, and B. Sen, *Anal. Chem.* **27,** 1460 (1955).
33. W. C. Tobie, *Ind. Eng. Chem., Anal. Ed.* **14,** 405 (1942).
34. H. E. Affsprung, and J. L. Robinson, *Anal. Chim. Acta* **37,** 81 (1967).
35. J. Studlar, and K. Janousek, *Coll. Czech. Chem. Comm.* **24,** 3799 (1959).
36. D. E. Jordan, and D. E. Mann, *Anal. Chim. Acta* **37,** 42 (1967).
37. A. K. Gupta, and J. E. Powell, *Talanta* **11,** 1339 (1964).
38. K. L. Cheng, and B. L. Goydish, *Talanta* **13,** 1161 (1966).
39. R. Pribil, and J. Cuta, *Coll. Czech. Chem. Comm.* **16,** 391 (1951).
39a. P. F. S. Cartwright, *Analyst* **86,** 688, 692 (1961).
40. A. M. Escarrilla, *Talanta* **13,** 363 (1966).
41. F. H. Firsching, *Anal. Chem.* **32,** 1876 (1960).

CHAPTER

5

TITRIMETRY

5.1 COMPLEXOMETRIC TITRATIONS

Methods that employ EDTA, EGTA, and other aminopolycarboxylic acids now dominate the field of volumetric analysis, but because such titrants are very unselective the application of masking techniques to complexometric procedures has proved to be most fruitful.

As shown in Table 5.1, the stability constants of metal complexes with EDTA extend over a wide range so that some discrimination among groups of metals can be achieved by pH control. In acid solutions (pH 1–3) ter- and tetravalent cations can be titrated without interference from divalent cations. At pH 5 all divalent cations except those of the alkaline earths are also titrated, and the alkaline earths are included if the titration is carried out at pH 10. These divisions correspond to differences in stability constants of the complexes, namely, $\log K > 20$ for Bi^{3+}, Fe^{3+}, Ga^{3+}, In^{3+}, Th^{4+}, and other ter- and tetravalent cations, $\log K = 14$–18 for divalent transition metals, rare earths, and Al^{3+}, and $\log K = 8$–11 for the alkaline earths.

Much greater selectivity can be achieved by using masking agents, either to prevent interference by unwanted cations or to displace a cation from its EDTA complex, followed by back-titration of the liberated EDTA. A wide range of complex-forming reagents is available from which to form, with particular cations, complexes that are very stable, preferably soluble and colorless, and inactive towards EDTA and the indicators that are used. The purpose of this section is to outline some of the published titrimetric procedures which depend on the use of masking techniques and which, in this way, have allowed an extension of the possibilities of complexometry, especially in the routine analysis of alloys, ores, concentrates, and other systems containing mixtures of metals. Difficulties arise in masking cations that have small complex-forming tendencies or that give rise to very intensely colored complexes, and in the determination of chemically related cations, such as Co–Ni or Zn–Cd, because of the lack of specific masking agents.

A limiting factor in the choice of possible indicators to use in complexometric titrations is that the stability constant of the indicator with

Table 5.1. Stability Constants of Metal Complexes with EDTA[a]

Cation	Log K_{MHL}[b]	Log K_{ML}[c]	pK_a(ML)[d]	Cation	Log K_{MHL}	Log K_{ML}	pK_a(ML)
Ag^+	3.1	7.3		Mn^{2+}	6.9	13.9	
Al^{3+}	8.4	16.1	5.8[e]	Mo^{5+}		6.4	
Am^{3+}		18.2		Nd^{3+}	10.5	16.4	
Ba^{2+}	2.1	7.8		Ni^{2+}	11.6	18.6	
Be^{2+}		9.3		Pb^{2+}	10.6	18.0	
Bi^{3+}	13.9	22.8		Pd^{2+}		18.5	
Ca^{2+}	3.5	10.7		Pm^{3+}		16.8	
Cd^{2+}	9.1	16.5		Pr^{3+}		16.2	
Ce^{3+}		15.8		Pu^{3+}		18.1	
Cf^{3+}		19.1		Pu^{4+}		17.7	
Cm^{3+}		18.5		Pu^{6+}		16.4	
Co^{2+}	9.2	16.2		Sc^{3+}		23.1	10.5
Co^{3+}		36		Sm^{3+}		17.0	
Cr^{3+}	15	23	7.4	Sr^{2+}	2.3	8.6	
Cu^{2+}	11.5	18.8	11.5	Tb^{3+}		17.6	
Dy^{3+}		17.8		Th^{4+}		23.2	7.0
Er^{3+}		18.1		Ti^{3+}		19	
Eu^{3+}		16.9		Tl^{3+}	14.5	22.5	
Fe^{2+}	5.3	14.3	9.1	Tm^{3+}		19.0	
Fe^{3+}	16.2	25.1	7.5[f]	V^{2+}		12.7	
Ga^{3+}		20.3	5.9	V^{3+}		25.9	9.5
Gd^{3+}		17.0		VO^{2+}	11.4	18.0	
Hg^{2+}	14.6	21.9	9.1	VO_2^+	11.4	18.0	
Ho^{3+}		17.9		Y^{3+}		18.0	
In^{3+}		25.0	8.8	Yb^{3+}		19.1	
La^{3+}	8.1	15.1		Zn^{2+}		16.5	
Lu^{3+}		19.5		Zr^{4+}		19.4	6.2
Mg^{2+}	2.3	8.9					

[a] Based on values in L. G. Sillén and A. E. Martell, "Stability Constants of Metal-Ion Complexes," *Chem. Soc. (London), Spec. Publ.* **17,** 634 1964. For 20–25° and an ionic strength of about 0.1.

[b] Log K for $M + HL \rightleftharpoons MHL$.

[c] Log K for $M + L \rightleftharpoons ML$.

[d] pK_a for $ML \rightleftharpoons M(OH)L + H^+$.

[e] Further pK_a at 10.0.

[f] Further pK_a at 9.5.

any kind of metal ion present must not be so high that the indicator becomes "blocked." Thus, in titrations of Mn and Zn against eriochrome black T, traces of Co, Ni, Cu, or Fe "block" this indicator. This behavior is strictly analogous to masking, and the indicator can be demasked if a suitable reagent is added which forms an even stronger complex with the interfering cation. In this particular example deblocking of eriochrome black T can be achieved in the back-titration of Co, Fe,

Cu, Ni, and Al by adding water-miscible organic solvents such as ethanol to diminish the stability constants of the indicator-metal complexes.[1]

ALUMINUM

Direct titrations are usually not very satisfactory, because reaction with EDTA is slow about pH 3.5 (unless the solution is heated) and equilibrium is unfavorable at lower pH values. Sajo[2] determined the total Al + Fe + Ti content of a solution by back-titration of excess EDTA with standard Zn at a pH less than 6.7 (using ferro–ferricyanide, benzidine indicator), then added phosphate to mask Ti (titrating the liberated EDTA with Zn), followed by sodium fluoride to mask Al. After heating and cooling the solution, the liberated EDTA was titrated to obtain the Al content of the solution. In this method Co, Th, Zr, La, and Ce interfere but Co can be masked by hydrogen peroxide.[2]

Masking by fluoride ion is a reasonably selective method for Al, although Ti and Zr are also masked. It has been used in the complexometric titration of Al–Ni–Zr alloys. Zirconium is titrated with EDTA in $2N$ sulfuric acid (against xylenol orange), whereas for Ni the titration is at pH 7 after adding sodium fluoride to mask Al and Zr.[3] For the analysis of an Al–In–Ni alloy, the total metal content was obtained by back-titration, with Zn, of excess EDTA at pH 5–5.5. By adding ammonium fluoride, boiling, and cooling, Al was displaced and the free EDTA was again titrated with Zn. The Ni content was determined on another portion by masking Al with triethanolamine, In with mercaptoacetic acid, and back-titrating with Ca in an ammoniacal solution after adding excess EDTA. Indium was determined on another aliquot by back-titration of excess CDTA with Ca, in ammoniacal solution, masking Al with triethanolamine and Ni with cyanide ion.[4]

Aluminum has been determined by adding excess CDTA to an acid solution, adjusting to pH 5–5.5 with hexamethylenetetramine, and back-titrating with lead nitrate, using xylenol orange. (This step leads to the titration of many other cations as well, including Fe(III).) Addition of ammonium fluoride then displaces Al from Al–CDTA, so that further titration gives the Al content.[5] This method has been extended to the analysis of systems containing Mg, Ca, and Fe as well as Al. Calcium is determined by titration with EDTA, in the presence of alkaline triethanolamine as a masking agent for Al and Fe, while Mg is precipitated as the hydroxide. The amount of Mg + Ca is obtained by a similar titration in ammonia buffer.[6] A modification enables Al and Cr to be determined in the presence of chromate.[7] HEDTA has been suggested for the estimation of Al and Ni in the presence of Mn. The method uses excess of HEDTA, buffering with acetate and boiling. Back-titration at room temperature with Cu against methylcalcein gives Ni + Al. Masking of Al with ammonium fluoride gives Ni, and hence Al by difference. Finally, Mn is titrated with EDTA at pH 9.5, in the presence of potassium tartrate.[8] In general, however, if many other metals are present some prior form of separation, such as their precipitation by adding strong alkali (which retains Al in solution as aluminate), is desirable.

In a mixture of Cr + Al, Cr(III) has been masked by oxidation with hydrogen peroxide in boiling alkaline solution, After EDTA titration of Al at pH 10.5 (against chromogen dark blue), Cr was determined by acidification with sulfuric acid, addition of potassium iodide, and titration of iodine with sodium thiosulfate.[9]

Iron and Al can be determined in admixture. Titration with EDTA at pH 1–2 in the presence of sulfosalicylic acid as indicator gives Fe(III). After adjusting to pH 5 and adding excess EDTA, Al is determined by back-titration of EDTA with Fe.

Hydrolytic masking of Al provides a selective method for the EDTA titration of Al and Ga. (See *Gallium*.) Other reagents that have been used to mask Al in EDTA titrations include borofluoride, tartrate and catechol-3,5-disulfonate ions, and acetylacetone. Citrate and sulfosalicylate are also suitable if only small amounts of Al are present.

ANTIMONY

Binary alloys of Sb with Zn, Sn, and Bi have been analyzed complexometrically by adding excess EDTA to an acid solution, adding a known amount of Cu(II) and titrating with EDTA at pH 2.5–3 [against 4-(2-pyridylazo)resorcinol] to determine the total metal content. Antimony is then masked by heating with tartaric acid and the liberated EDTA is titrated with Cu.[10]

Dimercaptopropanol, unithiol, citrate, fluoride, and oxalate have also been used as masking agents for Sb in complexometric titrations.

BARIUM

The stability constant of the Ba–EDTA complex is low so that its complexometric determination in the presence of other cations depends on displacement reactions. For the estimation of Ba in a Ba–Ca–Mg mixture, EDTA titration is carried out in dilute alkali (using thymolphthalexone as indicator), then a duplicate titration is carried out after adding ammonium sulfate to mask Ba as its sulfate. (The logarithm of the solubility product of $BaSO_4$ at an ionic strength of 0.1 is —9.2; compare —3.8 for $CaSO_4$ and —5.8 for $SrSO_4$).[11]

Similarly, in binary mixtures of Ba with Ca, Mg, Mn, Zn, or Cd, direct EDTA titration in ammonia/ammonium chloride buffers gives the total cation concentration. Addition of a known amount of magnesium sulfate then displaces the Ba from Ba–EDTA, and precipitates it as barium sulfate, so that subsequent EDTA titration enables the amount of Mg equivalent to the original Ba to be calculated. (When Mn is present, triethanolamine is needed to prevent its hydrolysis, and ascorbic acid to prevent its oxidation to Mn(III).)[12]

Fluoride ion is an alternative masking agent for Ba in such titrations.

BERYLLIUM

Because of its strong tendency to hydrolyze, Be cannot satisfactorily be titrated directly with complexometric reagents. On the contrary, Be can be precipitated as the hydroxide or as a hexamminecobalt(II) complex in the presence of EDTA as a masking agent for other cations. Conversely, to enable other metal ions to be titrated, Be has been masked with citrate, fluoride, or acetylacetone.

BISMUTH

In the direct complexometric titration of Bi at pH 1–3, using methylthymol blue, pyrocatechol violet, or xylenol orange as indicators, the main interference comes from ter- and quadrivalent cations. Iron (and Hg) can be reduced by ascorbic acid, and fluoride ion is a good masking agent for most of the other cations, including Sn.[13] Lead, Zn, and Cd do not interfere. Potassium thiocyanate can also be used to mask Hg, which can subsequently be demasked by silver nitrate.[14]

An analysis of a Bi–Cd–Pb–In mixture depended on back-titration, at pH 5–5.5, of EDTA with Zn (methylthymol blue) to give the total metal present, followed by a duplicate titration in the presence of mercaptoacetic acid as masking agent for Bi, so that the amount of Bi was obtained by difference. The sum of Bi + In was found by back-titration at pH 2.2–2.5 of an excess of EDTA, using Bi. Back-titration with Mn, in ammoniacal solution, in the presence of ascorbic acid and cyanide ion (to mask Bi and Cd, respectively) gave Pb + In.[15] Alternatively, Cd can be masked with 1,10-phenanthroline at pH 5–5.5. If necessary, Bi can be separated from other cations by precipitation from ammoniacal solution with sodium diethyldithiocarbamate in the presence of EDTA and potassium cyanide, followed by extraction into chloroform and digestion with nitric acid.[16] In the complexometric titration of Bi, Th can be masked by sulfate ions.[17]

Other masking agents for Bi in EDTA titrations include dimercapto-propanol, unithiol, mercaptosuccinate, and (if only small amounts of Bi are present) citrate ion. Alternatively, Bi can be hydrolyzed in a chloride-containing solution to form sparingly soluble bismuth oxychloride.

CADMIUM

Cadmium can be determined by direct titration with EDTA in ammoniacal solution (against thymolphthalexone or methylthymol blue) or in pH 5–5.5 buffer (against xylenol orange or methylthymol blue), but masking is necessary to improve selectivity. Masking agents that have been used include thiourea, sodium thiosulfate, or ascorbic acid and potassium iodide, to mask Cu; boiling with ammonium fluoride to mask Al, Ca, and Mg; acetylacetone or potassium fluoride (to mask Fe in acid solution); and triethanolamine (to mask Al and Fe in alkaline medium).

Because of their similarities, determination of Cd and Zn in the presence of each other is important. Most methods use the selective masking of Cd. An early method was based on the titration of Cd + Zn in ammonia/ammonium chloride buffer, using EDTA, followed by addition of sodium diethyldithiocarbamate which masked Cd by precipitation as $Cd(DDC)_2$, the liberated EDTA being titrated with Mg.[18] Extension to mixtures such as Cd–Mg–Ni–Zn and Cd–Pb–Zn is possible by using further masking agents such as potassium cyanide or thioglycolic acid, and demasking $Zn(CN)_4^{2-}$ with formaldehyde or chloral hydrate. Cadmium, but not Zn, displaces Pb from Pb–EGTA in the presence of sulfate.[19] Similarly, β-mercaptopropionic acid masks Cd but not Zn against titration with TTHA at pH 5 (urotropine buffer).[20] Copper is also masked as a colorless Cu(I) complex and does not interfere. In determining Zn by back-titration after masking with β-mercaptopropionic acid, a good end point is obtained if CDTA, but not EDTA, is used.[20] Dimercaptosuccinic acid has been used for the same purpose in EDTA titrations of Zn and Cd,[21] but is slightly less effective.[20] It masks Cd, Cu, and Hg. With Fe(III), Co and Ni it gives colored complexes: that with Fe(III) is dark red and interferes in the titration, even if triethanolamine is present. At pH 10, Pb, Cu, Ni, Co, and Cd are masked, but not the alkaline earth metals.

A large excess of potassium iodide masks Cd in the titration of Zn with EDTA (or, better, DTPA) at pH 5.[22] Copper and Hg are also masked; sulfosalicylic acid or pyrocatechol-3,5-disulfonate suppress Al. The reaction can also be used to displace Cd from Cd–EDTA, the liberated EDTA being titrated with standard Zn at pH 5–5.5.

Cadmium can be titrated directly in the presence of Pb and Zn by back-titration of excess EGTA in sodium hydroxide solution with Ca, using metalphthalein as indicator, Zn being masked as zincate: Pb is also masked by hydroxide ion. If necessary, Fe and Al can be masked by triethanolamine. As a check that the titration is of Cd, and not Cd + Ca, the Cd can be masked by potassium cyanide and the liberated EGTA titrated with Ca.[23]

For determining Cd in the presence of much Cu both can be masked by careful addition of potassium cyanide, followed by addition of Mg–EDTA which demasks Cd but not Cu, and the displaced Mg is titrated directly with EDTA, using eriochrome black T.[24]

If necessary, Cd can be separated from Zn and other metal ions by precipitation as its 1,10-phenanthroline iodide,[22] or as its sulfide by homogeneous precipitation using thiourea in sodium hydroxide. Dimercaptopropanol, cysteine, unithiol, tetren, and mercaptoacetate have also been used to mask Cd in complexometric titrations.

CALCIUM

Eriochrome black T, which is commonly used as an indicator in the complexometric titration of Ca, is easily blocked by traces of metal ions such as Cu, Co, Ni, Al, and Fe: when these are present, masking agents, such as sulfide for Cu, or cyanide for Cu and Ni, must be added. (In this case triethanolamine is not effective enough as a masking agent for Fe.) If acid alizarine black SN is used (in the absence of Mg), or calcon (if Mg is present),[25] the following masking agents can be added to a solution at pH 12 that is to be titrated with EDTA: triethanolamine for Al, Bi, Fe, Ti, and small amounts of Mn; cyanide ion for Cd, Co, Cu, Hg, Ni, Pt, Zn; dimercaptopropanol for Bi, Cd, Hg, Pb, Zn: unithiol for As, Bi, Cd, Hg, Pb, Sn, Zn[26]; and hydroxide ion masks Mg, Al, and Zn. β-Mercaptopropionic acid satisfactorily masks Fe, Bi, Co, Cu, and Hg (as colorless or slightly colored complexes) in the complexometric determination of Ca, Mg, Mn, and Ni.[27] Tartrate ion can, in many cases, be used instead of, or together with, triethanolamine. Sodium gluconate has been suggested for masking trivalent cations.[28]

In potassium hydroxide solutions Mg is masked by precipitation as $Mg(OH)_2$ (without removal from the system), so that Ca can be titrated directly with EDTA using murexide or calcein as indicator. If much

Mg is present, however, some Ca tends to be coprecipitated. This difficulty can be avoided by titrating both the Mg and the Ca with EDTA (at a somewhat lower pH), then making the solution strongly alkaline and displacing the Mg from the Mg–EDTA complex by back-titration with Ca.[29] Similarly, after titration with EDTA, Zn in ammoniacal solution can be displaced from its EDTA complex by Ca, then precipitated, for example, as its sulfide.[30]

At pH 13, with EGTA as titrant and calcon as indicator, Mg can be masked by tartrate ion in the complexometric determination of Ca.[31] The titration has also been carried out at pH 9.5–10, using zincon and a small amount of Zn–EGTA complex. The Ca displaces Zn from the complex, giving a blue Zn–zincon complex which persists until the end point is reached in the EGTA titration, whereupon the orange-yellow zincon ion is reformed.[32] The reason that Ca can be titrated in the presence of Mg is that the stability constants of their EGTA complexes differ markedly: log K for Mg–EGTA is 5.2, whereas log K for Ca–EGTA is 11.

An alternative approach to the determination of Ca in the presence of Mg is to obtain the total Ca + Mg by direct EDTA titration in ammonia buffer, and then to determine Mg under conditions where Ca is masked. This has been done by masking Ca with excess EGTA while Mg was titrated with CDTA using methylthymol blue.[33] Before making the solution alkaline, triethanolamine was added to mask Al and Fe, and thioglycolic acid to mask Cd, Cu, Sn, and Zn. Potassium cyanide in the alkaline solution masked Co and Ni. It has also been reported that in a mixture of Ca and Mg, Ca can be precipitated as molybdate from neutral solution (50% aqueous ethanol) and Mg can then be titrated with EDTA at pH 10–11 without removal of the precipitate.[34]

Many methods have been devised for the titrimetric analysis of mixtures of metal ions including Ca. Usually, determination of Ca is by difference between titrations in acid and ammoniacal solutions. Because Mn is titrated in neutral or alkaline solution, it would also be included with Ca. If present, it can be detected, and separately determined, by masking Ca and Mg as their fluorides.

In the analysis of a mixture such as Ca–Mg–Pb–Zn the total metals can be determined by back-titration of excess EDTA with Mg. Addition of cyanide ion to mask Zn enables Zn to be obtained by difference. Further addition of thioglycolic acid gives Pb. Calcium and Mg can be determined separately as described above. Masking with cyanide ion also makes it possible to determine Ca or Mg in admixture with Ni or Cu.

Consecutive titrations have been used to determine Ca, Mg, and Fe in the same aliquot. After the initial titration of Ca at a pH above 13 with EGTA (Mg masked as hydroxide; Fe, Al, Mn masked by triethanolamine), Mg was titrated with EDTA at pH about 11, and, finally, after destroying the triethanolamine with formaldehyde, the demasked Fe was oxidized to Fe(III) and titrated with EDTA.[35]

At pH 10, 2,3-dimercaptopropionic acid is a suitable masking agent for Pb, Cu, Ni, Co, and Cd, in the presence of Ca, but the Cu complex is colored and Fe(III) must be reduced to Fe(II).[36]

When masking is inadequate for titrations, recourse must be had to methods of separation, such as precipitation of Ca at pH 5–6 as the oxalate (with EDTA as a general masking agent), precipitation of the interfering cations, or use of ion-exchange separations in the presence of complexing species.

CERIUM

See *Lanthanum.*

CHROMIUM

Although the stability constant of Cr–EDTA is high, the hydrated Cr(III) ion reacts only slowly in slightly acid solution, while the resulting complex is so strongly colored that only small amounts of Cr can be determined by complexometric titration. The slowness of this reaction can be turned to advantage by carrying out a direct EDTA titration at pH 1–2 at room temperature (to determine the amount of interfering cations), then adding excess EDTA, boiling to react the Cr, and back-titrating with bismuth nitrate, using xylenol orange.[37] Similarly, in the analysis of Cr–Ni–Fe alloys, the sum of Fe + Ni was determined by back-titration of excess EDTA with lead nitrate at pH 5–6, using xylenol orange, then further EDTA was added, the pH adjusted to 1–2 and the solution boiled for 15 min. On cooling and re-adjusting to pH 5–6 with hexamethylenetetramine, the titration was repeated to give Cr.[38] In a modified procedure, a Cr–Ni–Fe mixture was boiled with excess EDTA at pH 1, triethanolamine was added to mask Fe and the solution was made alkaline. Titration with Ca against fluorexon gave Cr + Ni, and subsequent addition of potassium cyanide masked Ni which could be determined from the liberated EDTA.[38] Masking has also been used in the complexometric titration of Cr and Zn in their mixtures.[39]

Alternatively, Cr can be determined in alkaline solutions (where partially hydrolyzed complexes are formed) by back-titration with Mn, Zn, Ni, or Ca, in the presence of masking agents such as cyanide ion (if Mn or Ca is used as titrant). Chromium can also be masked by triethanolamine, or by reduction with ascorbic acid, or advantage can be taken of the slowness of formation and dissociation of Cr–EDTA; all of these have been suggested for determining Cr, Ca, Mn, and Ni in admixture.[40] Other possibilities as masking agents are pyrophosphate, or oxidation to chromate (so long as the indicator used in the complexometric titration is not, in turn, oxidized by chromate ion).

COBALT

Cobalt(III) forms a very much more stable complex with EDTA than does Co(II), and in slightly acid solutions, Co(II)–EDTA is very easily oxidized by Ce(IV), H_2O_2, or $KMnO_4$ to the intense red Co(III)–EDTA. In alkaline solution Co(II)–EDTA is oxidized by hydrogen peroxide to a stable blue peroxy complex which does not react with cyanide ion: this makes it possible to determine Co and Ni in the same solution. After EDTA titration at pH 5 to give Co + Ni, ammonia and hydrogen peroxide are added, followed by cyanide ion which displaces Ni from Ni–EDTA, giving $Ni(CN)_4^{2-}$. The liberated EDTA is titrated with Mg to give the total Ni.[41] The reaction can be used in more complicated systems, for example, for the determination of Co in a mixture of Al, Fe, Ni, and Ti, by masking Al, Fe, and Ti with fluoride ion.

In general, however, direct titration of Co(II) with EDTA is difficult if many other cations are present. Ammonium fluoride can be used to mask Al and thiourea to mask Cu. Potassium fluoride (made in solution by adding potassium carbonate to ammonium fluoride, because potassium fluoride in not sufficiently soluble) precipitates Fe as K_4FeF_6 at pH 5–5.5 and also masks Al, Ti, and the rare earths. In alkaline medium thioglycolic acid is useful for masking Ag, Bi, Cd, Cu, Pb, and Zn, but triethanolamine can only be used to mask Fe and Al as a displacement reaction (ie, from their EDTA complexes) because EDTA does not quantitatively displace Co from its triethanolamine complex (some of the Co(II) is oxidized to give the much more stable Co(III) complex). Dihydroxyethylglycine masks Bi, Cu, and Ti.

For the titration of Co(II) with EDTA ammonium thiocyanate can be used as indicator, the end point being given by the disappearance of the color of the thiocyanate complex when all the Co is masked by the EDTA. Alternatively, excess EDTA can be back-titrated with Ca in alkaline medium, using thymolphthalexone, methylthymol blue, or calcein, or with Pb at pH 5–5.5, using xylenol orange. Cobalt has been determined in alkaline ammonia solution, in the presence of Fe, Nb, Ta, Ti, and W (all masked by tartrate and fluoride ion) by back-titration of excess EDTA with standard Cu, using PAN [1-(2-pyridylazo)-2-naphthol].[42]

Masking agents for Co in EDTA titrations include cyanide ion, 1,10-phenanthroline and tetren, or if only small amounts of Co are present, mercaptoacetate and dimercaptopropanol.

COPPER

Copper(II) can be titrated directly with EDTA in acid, neutral or alkaline solution. Schwarzenbach's original procedure used an ammoniacal solution (murexide as indicator); at the other extreme Busev carried out the titration at pH 2.8 (against 7-(2-pyridylazo)-8-hydroxyquinoline).[43]

Use has been made of the fact that Cu can be masked by thioglycolic acid in alkaline medium, giving a colorless complex, so that Cu is displaced from Cu–EDTA, whereas Ni–EDTA and Co–EDTA are stable. This provides a basis for the analysis of Cu–Ni and Cu–Co mixtures by back-titration of EDTA with Ca, using methylthymol blue or thymolphthalexone as indicator, in the absence, and then in the presence, of thioglycolic acid.[44] The method can also be used for Cu + Mn.

In acid solution thiourea masks Cu against EDTA, so that Cu can be determined in the presence of Al, Co, Ni, or Pb by back-titration of excess EDTA at pH 5–5.5 with Pb (using xylenol orange) in the absence and in the presence of thiourea. Similarly, Cu and Zn (or Cd) have been titrated directly, the Zn being determined in acid solution after masking Cu with sodium thiosulfate,[45] or in ammoniacal solution (murexide) after adding cyanide ion and demasking the zinc cyanide complex.[46] It has also been reported that Zn can be demasked by acidifying the cyanide solution with hydrochloric acid, followed by immediate addition of acetate buffer (pH 5.7–5.9), and direct titration with EDTA, using xylenol orange.[47]

Copper can be determined in a mixture of Co, Cu, Fe, and Ni by back-titration of a known excess of EDTA with Pb, followed by a duplicate titration in which thiourea is added to mask Cu, and also some fluoride ion to prevent the reduction of Fe(III) to Fe(II).[48] An extension of this approach enables a mixture containing Cu, Ni, Pb, and Zn to be analyzed using four titrations with EDTA. The first gives the total metal content, in the second titration Cu is masked by thiourea, in the third Pb, Zn, and Cu are masked by mercaptoacetic acid in ammonia buffer, and in the fourth, Cu, Ni, and Zn are masked by cyanide ion in ammonia buffer.[49]

Triethanolamine cannot be used to mask alkaline solutions of Fe and Al in cases where Cu is determined by EDTA titration, because the Cu–triethanolamine complex is too stable.

In the titration of Cu and Ni with EDTA, at pH 3–4 in hot solution and using PAN as indicator, interference by tungsten can be masked by tartrate, fluoride, or phosphate ions, but Mo is only partially masked: CDTA is a better titrant.[50]

Many other possibilities exist for masking Cu in complexometric titrations. Ascorbic acid, hydroxylamine hydrochloride, cysteine, iodide ion, thiocyanate ion, or thiosulfate ion reduces Cu(II) to Cu(I) which does not react with EDTA. Precipitation as copper sulfide, and complex formation with trien, tetren, 3-mercapto-1,2-dihydroxypropane, 1,10-phenanthroline, and thiocarbohydrazide have also been used. When only small amounts of Cu are present, dimercaptopropanol, sodium diethyldithiocarbamate and thiosemicarbazide are also suitable.

GALLIUM

The borofluoride ion has been used to mask Al in the EDTA titration of Ga in acetate buffer (pH 3.8), with morin as indicator.[51] Fluoride ion is also suitable as a masking agent for Al when the hot solution is titrated at pH 1.6–2.0 with Cu–EDTA, using PAN as indicator.[52]

In the presence of boiling 25% ammonium chloride solution at pH 1.3–1.4 a large amount of In can be masked when Ga is titrated with EDTA, against xylenol orange: ascorbic acid masks Cu and Fe: iodide ion masks Cd and Zn.[53] Sodium borate and sodium fluoride have been used to mask Ti and Be at pH 2.[54]

The EDTA titration of Ga can be made more selective by titrating all metals at pH 6.5 (back-titration of EDTA with Pb, using (2-pyridylazo)resorcinol as indicator, in the presence of tartrate as masking agent), then adjusting to pH 10.5 to demask Ga (as the hydroxide) in the presence of excess Pb, and titrating the remaining Pb with EDTA.[55] A modification, in which the first solution is titrated at pH 5.5 after boiling, and the second solution (as a separate aliquot) is treated with excess EDTA and tartaric acid solution before being back-titrated at pH 10.5 with standard Cu, enables Al to be determined.[55]

Citrate ion has also been used to mask Ga in complexometric titrations.

If masking is inadequate, Ga can be separated from interfering elements by extraction as its chloro complex into isopropyl ether from 7*M* hydrochloric acid.

GERMANIUM

Probably because of the ease with which Ge can be separated from other elements by distillation as $GeCl_4$, or by extraction of $GeCl_4$ into carbon tetrachloride, masking procedures have not been described for use in the complexometric determination of Ge by back-titration of excess EDTA.

GOLD

Gold has been determined by indirect complexometric titration, based on displacement of nickel from potassium nickel cyanide in slightly alkaline solution. Addition of excess EDTA was followed by back-titration with Mn(II), against eriochrome black T.[56] Palladium can be determined in the same way.[56]

HAFNIUM

See *Zirconium.*

INDIUM

The high stability constant of In–EDTA makes it possible to titrate In in acid solution, thereby avoiding interference from many other metal ions. Analysis of an Al–In–Ni mixture has already been discussed (see *Aluminum*). Titration of a Bi–Cd–Pb–In mixture was also discussed under *Bismuth.* Similarly, for In–Ni–Zn, In has been determined by back-titration of excess CDTA with Ca in the presence of cyanide ion to mask Ni and Zn, the total In + Ni + Zn being obtained in a similar titration without cyanide ion. Back-titration of EDTA with Ca in the presence of mercaptoacetic acid (to mask In and Zn) afforded Ni.[4]

In ammoniacal tartrate solution, of about pH 8, In and Zn can be determined by back-titration with standard In solution after adding excess EDTA. (The tartrate masks In against precipitation). Addition of cyanide ion displaces EDTA from Zn–EDTA, and this is also titrated with In.[58] Alternatively, two titrations can be carried out, one of these being direct, after adding cyanide to mask Zn, the other being a back-titration of excess EDTA with Hg(II), to an eriochrome black T end point.[59]

The determination of In in the presence of Zr is possible by masking the latter with fluoride ion in an EDTA titration at pH 3: the rare earth metal fluorides are also precipitated. Back-titration with Fe(III) at pH 2–2.5 after adding excess EDTA to a duplicate solution, without fluoride ion, gives In + Zr.[60]

Indium has been determined by EDTA titration of a boiling solution at pH 8–10, in the presence of tartaric acid. Cyanide ion masked Cd, Co, Cu, Hg, Ni, and Zn, while the presence of hydroxylamine reduced Fe(III) to Fe(II), which was masked by forming ferrocyanide ion.[61]

IRON

The ease with which Fe^{3+} undergoes hydrolysis, except in strongly acid solutions, and the yellow color of the Fe(III)–EDTA complex present difficulties in the complexometric titration of Fe. One way of avoiding these is to reduce Fe(III) to Fe(II) with ascorbic acid and titrate hot with EDTA in a hexamethylenetetramine buffer (pH 5–5.5) against methylthymol blue, masking Al with fluoride ion.[62] Iron and Al have been determined in copper alloys by back-titration of excess CDTA at pH 5–5.5 with Pb (xylenol orange) in the presence of thiourea to mask Cu: a duplicate titration in which ammonium fluoride is added before the CDTA (to mask Al) gives Fe. In the presence of Mn, Sn, Pb, Ni, and Zn, Fe can still be determined by difference, using a back-titration after masking the Fe by precipitation with potassium fluoride (see *Aluminum*).[63] Similarly, in the titration of Fe(III) at pH 1 with EDTA, Al can be masked by ammonium fluoride or salicylic acid, and Cu with thiourea.[64] Tungsten is masked by tartrate. Nickel and Co have also been masked by tartrate when Fe(III) was titrated with EDTA at pH 1.0–1.3, using *N*-phenylbenzohydroxamic acid as indicator.[65]

Many reagents are available for masking Fe in such titrations. They include cyanide ion (usually in the presence of reducing agents), citrate, tartrate, pyrophosphate, mercaptoacetate, and mercaptosuccinate, sulfide ion, and triethanolamine. It is sometimes adequate to reduce Fe(III) to Fe(II) with ascorbic acid, hydroxylamine, hydrazine, or stannous chloride.

LANTHANUM, CERIUM, AND THE RARE EARTHS

The stability constants of the complexes of rare earth cations with aminopolycarboxylic acids increase more or less progressively from La to Lu, the greatest differences being found with CDTA ($\log K = 16.3$ for La–CDTA, 21.5 for Lu–DCTA), EDTA (15.5 and 19.8), and propylenediaminetetraacetic acid (16.4–20.6). Although these differences are important for fractional separations, they are not great enough to enable individual elements to be determined. The constants are large enough, however, to make it easy to obtain the sum of the rare earths present. The first complexometric titrations of rare earths were at pH 8–9 in the presence of tartrate or citrate ions to prevent hydrolysis and precipitation,[66] but it is easier to use solutions at pH 5–6, with xylenol orange or methylthymol blue as indicators.

Back-titration of excess EDTA at pH 5–6 with Zn avoids difficulties in titrating rare earths in the presence of oxalates.[67] If the titration is carried out in weakly alkaline solutions, Cu, Cd, and Zn can be masked by potassium cyanide[66]; in slightly acid solutions 1,10-phenanthroline is suitable.[68] At pH 7–8, Al and Fe are masked by sulfosalicylic acid, while thiourea selectively masks Cu.[69] Sulfosalicylic acid masks Th at pH 5.[69] Similarly, by using DTPA instead of EDTA interference by phosphate can be overcome.[69a] At pH 10, unithiol has been proposed for masking Zn, Cd, Hg, Pb, Sn, Bi, Ga, and In in the EDTA titration of the rare earths (by back-titration with Mg, using eriochrome black T as indicator).[70]

In acid solutions the alkaline earth cations do not interfere. Conversely, in stronger acid, the rare earths are not titrated. Thus Bi has been titrated at pH 2, followed by Nd or Pr at pH 5–6[71]; or again Sc at pH 1.8–2.2, followed by rare earths at a higher pH.[72] Similarly with DTPA or TTHA, Th (and Sc) are determined at pH 2.5–3.5, followed by the rare earths at pH 5–6.[73] This pH difference in stability has been used to determine rare earths by adding EDTA, titrating the excess at pH 2 with ferric chloride, using sulfosalicylic acid as indicator, then adjusting to pH 4.5–6, and determining the rare earths by EDTA titration against xylenol orange.[74]

Selectivity in the titration of rare earths can be achieved by masking them with fluoride ion, thereby displacing them from their EDTA complexes. This avoids the necessity of separating them from most of the transitional metal ions. Back-titration with Zn(II) solution, using xylenol orange as indicator, after adding excess EDTA, gives the total metal content, including heavy metals and iron. Subsequent addition of ammonium fluoride liberates EDTA from the rare earth complexes, the amount being determined by further titration with Zn(II). Interfering cations such as those of Zr and Th can be determined in more acid solutions, also using fluoride ion in a displacement titration.

An alternative is to mask the rare earths with oxalate ion.

Acetylacetone has been used as a masking agent for Al in the titration of Ce(III) and La(III) at pH 5–5.5 with DTPA (using xylenol orange); ascorbic acid was added to reduce any Ce(IV).[75] In the EDTA titration of Lu in an Al–Lu alloy sulfosalicylic acid was added to mask Al.[76]

LEAD

The complexometric determination of Pb can be made much more selective by appropriate use of masking agents such as cyanide ion for Ag and many bivalent metal ions including Cd, Co, Cu, Hg, Ni, and Zn,[77] fluoride ion for Al, acetylacetone for traces of Fe, and triethanolamine for Al, Fe, Sn, Sb, while, at the same time, Pb is masked against precipitation as the hydroxide by adding tartrate or triethanolamine before making the solution alkaline. In this way, using an ammonia/ammonium chloride buffer of pH 10, and 4-(2-pyridylazo)resorcinol as indicator, Pb has been titrated with EDTA in the presence of Cu, Ni, Zn, Cd, Al, Fe, Sn, Sb, and Bi.[78] Mixtures of Pb with any one of these metals can readily be analyzed by complexometric titrations.

To determine Pb and Sn in tellurides, the sum of Pb + Sn was obtained by back-titration of EDTA at pH 4–5 with standard Pb. Addition of ammonium fluoride displaced Sn from its EDTA complex, and the free EDTA was again titrated. The lead content could also be determined on a separate aliquot in which Sn was masked with tartaric acid.[78a]

There is no highly selective masking agent for Pb in EDTA titrations, but reagents that are commonly used are sulfur-type ligands such as unithiol, dimercaptopropanol, sodium diethyldithiocarbamate, mercaptoacetate, and 3-mercaptopropionate. Alternatively, Pb can be precipitated with molybdate and sulfate ions.

MAGNESIUM

See also *Calcium*.

Magnesium is readily titrated with EDTA in ammoniacal solution, but in stronger alkali, such as $0.1M$ sodium hydroxide, Mg is precipitated as the hydroxide. This property has been used in a method for analyzing Mg in Al-base alloys. The sample was dissolved in sodium hydroxide in the presence of potassium cyanide and triethanolamine (thereby masking Al, Ni, Fe, and Mn), followed by hydrogen peroxide to mask Ti and Cr. Magnesium remained as an impure precipitate which, after washing, was dissolved in dilute hydrochloric acid. The solution was made alkaline with ammonia and titrated with EDTA, again in the presence of masking agents.[79] Magnesium can be titrated in the presence of calcium if the latter has been precipitated with oxalate ion, and in the presence of barium if the latter has been precipitated with chromate or, less effectively, sulfate.

The usual methods of masking Mg are by precipitation as the fluoride or the hydroxide.

MANGANESE

The usual conditions for the complexometric titration of Mn are an ammonia buffer at pH 10, the presence of a reducing agent such as ascorbic acid or hydroxylamine to prevent oxidation to Mn(III), and tartrate or triethanolamine to mask Mn against precipitation as the hydroxide.[80] Differences in the stability constants of their CDTA complexes enable mixtures such as Fe, Mn, and Mg to be titrated consecutively at pH 2, 6–6.5, and 11.2–11.3.[81] When masking agents are present, back-titrations, for example with Mg, are often useful.

1,10-Phenanthroline masks Co, Cu, and Ni in the EDTA titration of Mn at pH 8–9 in ammonium acetate buffer,[82] triethanolamine masks Fe and Al, and cyanide ion masks Cd, Co, Cu, Hg, Ni, and Zn. However, sufficient amounts of Fe persist in the solution that eriochrome black T may be blocked. To avoid this methylthmol blue or thymolphthalexone can be used, and in this way Mn can be determined in iron alloys:[83] Masking of iron by reduction with ascorbic acid and conversion to ferrocyanide in ammoniacal solution is also suitable, with the addition of pyrocatechol-3,5-disulfonate to mask Al and Ti.[84] The masking of iron by precipitation with potassium fluoride cannot be used if Mn is present because coprecipitation occurs.[85]

Thioglycolic acid masks Pb, Cu, Bi, and In, and cyanide ion can be used for Cu, Zn, Cd, Hg, and many of the transition metal ions, but a reducing agent should be present to prevent Mn from being oxidized to $Mn(CN)_6^{3-}$ and thereby masked. A mixture containing Cu, Ni, and Mn can be analyzed by back-titration of excess EDTA with Mg, alone and after addition of these complexing species. To determine Mn in the presence of Ca and Mg, it is necessary to titrate first with EDTA to give Mn + Ca + Mg, and then add ammonium fluoride, followed by back-titration of EDTA with an Mn(II) solution containing hydroxylamine. In this way CaF_2 and MgF_2 are precipitated without contamination by Mn. The back-titration gives Ca + Mg and hence, by difference, the Mn content.[86]

Alternatively, masking of Ca and Mg with ammonium borofluoride permits the direct titration of Mn with EDTA, using methylthymol blue. Addition of concentrated ammonia demasks the Ca and Mg from these complexes, and both cations can then be titrated.[87] For analyzing copper alloys Cu is masked by titrating with potassium iodide and sodium thiosulfate, Mn is then titrated with EDTA in ammoniacal solution, with cyanide present as masking agent and ascorbic acid to prevent oxidation.[88]

MERCURY

Because of the ease with which Hg forms hydrolyzed species, its conditional constant with EDTA is not high. Most of the complexometric titrations for Hg gain in selectivity by using a technique where the total cation concentration is determined, and then the Hg is masked, the displaced EDTA being titrated with a suitable cation. At pH 1 thiosemicarbazide or excess thiocyanate ion is suitable. (In fact, Hg(II) can be titrated with thiocyanate ion, using Fe^{3+} as indicator, without recourse to complexometry.)

At pH 5–6 Hg(II) can be masked by reduction to the metal with formic acid or ascorbic acid. The use of thiosemicarbazide to mask Hg enables the analysis of mixtures such as Hg–Cd, Hg–Pb, and Hg–Zn by complexometric titration: Ag and Cu are also masked. For the analysis of a Cu–Hg mixture, the total metal content has been determined by back-titration, with Cu, of excess EDTA (against murexide), followed by masking of Hg with potassium iodide and further titration with Cu.[89]

An acidimetric method has been proposed for Hg, based on the formation of hydroxide ions when Hg in Hg–EDTA is masked by thiosulfate ion in neutral solution[90]:

$$HgEDTA^{2-} + 2\,S_2O_3^{2-} + 2\,H_2O \rightarrow Hg(S_2O_3)_2^{2-} + H_2EDTA^{2-} + 2\,OH^-$$

Other masking agents for Hg include such sulfur-type ligands as dimercaptopropanol, mercaptoacetate, unithiol, thiourea, cysteine, and sodium diethyldithiocarbamate (for small amounts of Hg), as well as bromide, chloride and cyanide ions, trien, and tetren.

NICKEL

Nickel has been determined in the presence of Co by adding ammonia to the solution, then sufficient potassium cyanide and hydrogen peroxide to form $Co(CN)_6^{3-}$ and $Ni(CN)_4^{2-}$: careful addition of silver nitrate, to provide a slight but permanent turbidity, demasked the Ni which could then be titrated directly with EDTA, against murexide.[91] Formation of $Co(NH_3)_6^{3+}$ by oxidation of Co(II) with ammonium persulfate in the presence of strong ammonia has also been used to mask Co in the titration of Ni with EDTA.[92]

By making use of the masking of Cd, Cu, Pb, and Zn (but not of Ni) by dimercaptopropanol[93] or thioglycolic acid,[94] binary mixtures such as Ni–Cu, Ni–Zn, and Ni–Pb can be analyzed by EDTA titration. Inclusion of potassium cyanide extends possibilities to ternary mixtures such as Ni–Pb–Zn. In alloys containing Fe, Co, and Ni, triethanolamine in alkaline solution masks Fe and Al, and Co(II) is oxidized to Co(III) with hydrogen peroxide (and then forms a stable complex with triethanolamine). In mixtures such as Ni–Mg–Zn, Ni–Ca–Cd, or Ni–Pb–Mn, the Ni is usually masked by adding potassium cyanide and then carefully demasked by formaldehyde or chloral hydrate. In alkaline solutions triethanolamine is suitable as a masking agent for Al and traces of Mn; in weak acid Al and Fe can be masked by ammonium fluoride.

Nickel–CDTA in ammonia solution is stable to cyanide ion, whereas Cu, Zn, and Cd react quantitatively. Thus back-titration of CDTA with Mg (against eriochrome black T) makes it possible to analyze mixtures of Ni–Cu, Ni–Zn, and Ni–Cd.[87] In a mixture containing Ni, Al, and Mn back-titration of 2-hydroxyethylethylenediaminetriacetic acid at pH 5 with Cu (using methyl calcein or methyl calcein blue) gave Ni + Al. Addition of ammonium fluoride and boiling then displaced an amount of HEDTA equal to the Al content. Finally, after adjustment to pH 9.5, Mn was titrated with EDTA.[95]

In the EDTA titration of Ni and Cu (using murexide) tartrate masked Al, Fe(III), Ti, Cr(III), U, Ga, and other multivalent cations, triethanolamine with hydrogen peroxide masked Co, Fe, and Al, hydroxylamine hydrochloride masked Cu, and fluoride ion masked Ca, Mg, and Be.[96]

1,10-Phenanthroline and tetren have also been used to mask Ni in EDTA titrations.

PALLADIUM

See *Gold.*

POTASSIUM

A titrimetric method for potassium depends on an initial precipitation as potassium tetraphenylborate, $KB(C_6H_5)_4$, which is separated and dissolved in acetone. The tetraphenylborate ion displaces Hg from Hg–EDTA added in excess, and the EDTA is determined by back-titration with Zn.[97] In this method ammonia can be masked by formaldehyde.[98]

RARE EARTHS

See *Lanthanum.*

SCANDIUM

The stability constant of Sc–EDTA is so high that scandium can easily be titrated in acid solution (pH less than 3 to minimize hydrolysis of Sc^{3+}). Scandium can be determined, for example, in the presence of rare earths, the alkaline earths, and many bivalent metal ions. Masking of Cu (with thiourea) and Fe(III) (with reducing agents), narrows the main interferences in the titration to tetravalent ions such as those of Th, Ti, Zr, and Hf. If Th and Sc are precipitated as their hydroxides from a triethanolamine medium (which masks Al, Fe, and other metals), they can then be determined by titration at pH 2.5–3.5 with EDTA to give Sc + Th. Addition of excess TTHA and adjustment of the pH to 5–5.5 displaces Th from its EDTA complex to form Th–TTHA, whereas Sc–EDTA does not react. Titration with Zn of the excess EDTA (giving Zn–EDTA) and TTHA (giving Zn_2–TTHA) enables the amount of Th, and hence of Sc, to be calculated.[99]

Following titration with EDTA at pH 2.5–3.5, against xylenol orange, Th, Sc, and the light lanthanides can be displaced from their complexes by adding phosphate and titrating the liberated EDTA with Zn.[100,101] At pH 5–5.5, phosphate displaces Sc, but not Th, from its complex with TTHA (triethylenetetramine-*N,N,N′,N′,N″,N″*-hexaacetic acid). This forms the basis of a method for determining Sc and Th in the presence of each other.[102]

Scandium can be masked by fluoride or oxalate ions.

SODIUM

An indirect complexometric method for sodium is based on precipitation as its zinc uranyl acetate, followed by titration of the zinc with EDTA after dissolution in an ammonia buffer, using eriochrome black T as indicator. In this determination uranyl ion is masked by the addition of ammonium carbonate.[103]

STRONTIUM

Complexometric titration of strontium appears to have been very little studied; most methods for Ba or Ca would also include Sr. In the absence of suitable masking procedures, prior separation of Sr would usually be necessary, for example, by passage through an ion-exchange column.[104] Fluoride ion would appear to be a suitable masking agent for Sr.

THALLIUM

In the determination of Tl(III) by titration with EDTA, using methylthymol blue, alkaline tartrate (pH 7–10) has been used to mask Tl(III) against precipitation as the hydroxide: it also masks against reaction of Tl(III) with bromide or chloride ions.[105] The complexometric titration of Tl can be made highly selective by oxidizing Tl(I) with aqueous bromine in an acid solution containing excess EDTA, to form Tl(III)–EDTA. After bringing the solution to pH 9 with aqueous ammonia the excess EDTA is titrated with Pb, using 4-(2-pyridylazo)resorcinol as indicator. The Tl(III) is reduced to Tl(I) with sodium erythrobate (sodium araboascorbate), and the liberated EDTA is again titrated, to give the amount of Tl.[106] This method can be used, for example, in the presence of In, Mn, Cu, Fe, and Zn.

Other reagents that have been used to mask Tl(III) in EDTA titrations are mercaptoacetate and thiourea. Alternatively, it can be reduced to Tl(I) with cysteine, bisulfite, or formate and masked by iodide ion.

THORIUM

Because Th is usually titrated with EDTA around pH 2–3, few bivalent cations interfere. Reduction with ascorbic acid effectively masks Fe and Hg. In a mixture of Th and Zr the total metal content has been obtained by back-titration at pH 2.5 with Bi against xylenol orange. Acidification to pH 1 and addition of excess ammonium sulfate masked Th, and the displaced EDTA was again titrated.[107] Using TTHA or DTPA, Th can be titrated at pH 2.5, followed by the lanthanides at pH 5–5.5. No suitable masking agent is at present known for Sc in Th titrations, but mixtures can still be determined by differential titrimetry using TTHA and EDTA.[99] In the EDTA titration of Th, W can be masked by hydrogen peroxide.[108] Citrate or fluoride ion can be used to mask Th. Dimercaptopropanol masks Hg(II), Cu, and Pb in the EDTA titration of Th, and bishydroxyethyldithiocarbamate is a suitable masking agent for Hg(II), Pb, and Ni.[109]

TIN

The Sn(II) complex of EDTA is much more stable than the corresponding Sn(IV) complex. It is possible, for example, to titrate Sn(II) directly with EDTA at pH 5–6 (using methylthymol blue) while, at the same time, Sn(IV) is masked by fluoride and tartrate ions.[110] Under these conditions, Sb(III) is also masked by tartrate. White metal has been analyzed by back-titration of excess EDTA in hexamethylenetetramine buffer (pH 5–6) with Pb, using xylenol orange, after masking Cu in acid solution with thiourea. The amount of complexed EDTA gave Pb + Sn. Masking with fluoride ion then displaced Sn, which could be estimated from the EDTA liberated. A titration without thiourea gave Cu + Pb + Sn.[111]

Dimercaptopropanol, unithiol, dithio-oxalate, oxalate, citrate, and triethanolamine are masking agents for Sn, which can also be precipitated by hydroxyl ion as metastannic acid.

TITANIUM

The hydrated cation, Ti^{4+}, undergoes hydrolysis too easily for satisfactory titration with EDTA. A back-titration at pH 1–2 has been used, with bismuth nitrate as titrant and xylenol orange as indicator. Hydrogen peroxide masked Ti against precipitation, while interference by Al and Th was avoided by masking with salicylic acid and ammonium or sodium sulfate, respectively.[112] Titanium has also been determined by back-titration, using lactic acid to displace Ti from Ti–EDTA: this reaction does not occur if the peroxy–Ti–EDTA complex is formed.[113] For the titration of Al and Ti in ferrotitanium, Ti is displaced from Ti–EDTA by adding tartaric acid, and the liberated EDTA is titrated with Zn. Similarly, by adding fluoride ion, Al is removed from Al–EDTA.[114]

Masking agents for Ti that have been used in complexometric titrations include citrate, fluoride, phosphate and sulfate ions, triethanolamine and catechol-3,5-disulfonate.

URANIUM

Reduction of U(VI) with sodium dithionite in the presence of EDTA leads to the formation of U(IV)–EDTA, so that back-titration of the excess EDTA enables U to be determined.[115] Thorium and U(VI) have been determined in this way, the Th being titrated first at pH 3.5 with EDTA, using xylenol orange as indicator, followed by reduction of U(VI).[116]

More usually, however, the requirement is to mask UO_2^{2+} while other metals are titrated. This can be achieved by adding carbonate, citrate, phosphate, sulfosalicylate, tartrate, or fluoride ions, or by treatment with hydrogen peroxide.

VANADIUM

Vanadium and aluminum in steel have been determined by boiling the solution with an excess of EDTA and then back-titrating at pH 5.0–5.2 with Zn, in the presence of ascorbic acid as a masking agent. In a duplicate titration hydrogen peroxide was added after boiling and cooling the solution. This displaced V from its EDTA complex, masking it as a peroxy complex, so that only Al was titrated.[117]

YTTRIUM

In a mixture of Be and Y back-titration of excess EDTA with Zn, in acid solution, using eriochrome black T, gave the content of Y. When the solution was made alkaline, Be was precipitated as the hydroxide and determined gravimetrically: the EDTA masked Y against coprecipitation.[118]

ZINC

Conditions for titrating Zn are the same as for Cd (except that thymolphthalexone is blocked by Zn and cannot be used as an indicator: zincon or naphthol violet are alternatives). The stability constant of the Zn–EDTA complex is intermediate between those of the alkaline earths and the cations of Zr, Th, and Bi, so that in a mixture such as Bi–Ca–Zn titration at pH 1–2 gives Bi, at pH 5 Zn is titrated, and then, at pH 10, Ca is also determined. The difference in stability constants of the Zn and Ca complexes with DTPA is greater than for EDTA and facilitates the direct titration of Zn in the presence of Ca.[119]

Zinc can be determined in a mixture with Mn and the alkaline earths by titrating the total metals, then masking Zn by the addition of cyanide ion and titrating the liberated EDTA with Mg. Cyanide ion also masks Pd(II).[119a] Other displacement reactions that can be used in determining Zn include the addition of dimercaptopropanol, unithiol, or thioglycolic acid to displace Zn from Zn–EDTA without affecting Ni–EDTA or Co–EDTA. Cyanide ion displaces Zn from Zn–CDTA, whereas Ni–CDTA is stable.[120] Other examples have been discussed under *Demasking* (Chapter 4) and *Cadmium*.

In acid solutions sodium thiosulfate masks Cu, so that Zn can be titrated with EDTA (using xylenol orange) without interference from Cu.[121] Likewise, sodium thiosulfate masks Hg(II), as $Hg(S_2O_3)_2^{2-}$. In the EDTA titration of Zn, using dithizone as indicator, mercaptopropionic acid has been used to mask Pb,[122] and pyridinium hydroxyethyldithiocarbamate for both Cd and Pb at pH 4.8–5.0.[123] Zinc can also be titrated selectively with EDTA (using eriochrome black T or xylenol orange) at pH 6 in the presence of up to an equal amount of Cd if the latter is masked by 2,3-dimercaptosuccinic acid. Up to half as much Cu can also be masked in the same way, but all Fe present must be reduced to Fe(II).[124]

In iron alloys Zn has been titrated with EDTA at pH 8 (using eriochrome black T) in the presence of citrate ion to mask Fe.[125]

ZIRCONIUM

Titrimetric procedures for Zr also determine Hf so that, in analyses for Zr, the sum of Zr + Hf is ordinarily implied. In *M* nitric acid solution, Zr can be titrated directly with EDTA, against xylenol orange. This enables Zr to be determined in the presence of most cations but fluoride and sulfate ions interfere. In less acid solutions (pH above 1), fluoride ion can be masked by Be. Mixtures containing Th and Zr can be analyzed by back-titration of excess EDTA at pH 2.6–2.8 with bismuth nitrate (using xylenol orange), followed by masking of Th with sulfate and titration of the liberated EDTA at pH 1.2–1.3.[126] Copper can be masked by thiourea, and Fe by reduction to Fe(II).

Citrate, fluoride and tartrate ions can be used to mask Zr.

5.2 OTHER TITRATIONS

EDTA is also a useful masking agent in the more traditional kinds of titrations, such as the determination of Mn in Al alloys by iodometric titration of permanganate at pH 2.5–3, in which EDTA masks Cu and Fe.[127] In the argentometric titration of cyanide ion, using the Ag–bromopyrogallol red–1,10-phenanthroline mixed complex as indicator, EDTA masks a large number of cations.[128] It is also used to mask Cu, Pb, Zn, Cd, Ni, and Co when Ag is determined by precipitation with iodide ion, followed by potentiometric titration of the excess iodide ion with standard Hg(II), using Ag amalgam as an indicator electrode.[129] Similarly, in the potentiometric determination of Ag with dithiooxamide in potassium hydrogen phthalate buffer, EDTA and fluoride ion mask Fe, Co, Pb, and Cu.[130] EDTA masks Zn and Mn in the potentiometric titration of Ag with benzotriazole in acetate buffer.[131]

Arsenate can be determined as $MgNH_4AsO_4 \cdot 6H_2O$ by precipitation titration with magnesium sulfate in an ammonia/ammonium chloride buffer containing aqueous ethanol, if eriochrome black T is present to indicate the end point. Cyanide ion masks Fe and Cu after reduction with ascorbic acid, and EDTA masks the alkaline earths.[132] Alternatively, Fe(III) and Al are masked with triethanolamine if thymolphthalexon is used as indicator.[133] EDTA is also suitable for masking interfering metal ions in the alkalimetric titration of borate as its mannitol complex.[134]

Complex formation can affect the oxidation–reduction potentials of metal ions,[135] because of differences in the stability constants of the complexes formed by the oxidized and the reduced forms of the metal ion concerned. In the system Fe(II)/Fe(III) anionic ligands such as phosphate, EDTA, or fluoride ion form much more stable complexes with Fe(III) than with Fe(II), so that oxidation of Fe(II) to Fe(III) is facilitated. The use of EDTA masks Fe(III) against its reduction by many of the common reducing agents. Similarly, although Fe^{3+} is readily reduced by iodide ion, to give Fe^{2+} and iodine, addition of 0.1 *M* fluoride ion to a weakly acid solution of Fe(III) masks this reaction completely. In fact, in the presence of EDTA, Fe^{2+} is oxidized by iodine to give Fe^{3+} (as its EDTA complex) and iodide ion. Ligands such as 2,2′-bipyridine and 1,10-phenanthroline preferentially stabilize Fe(II) and Co(III), so that the titration of Co(II) with Fe(III) in the presence of these complexing species is analytically feasible. Addition of fluoride ion to Ti(III) to form TiF_6^{3-} masks it against oxidation by iodine.

In titrations involving oxidation or reduction, interfering reactions can sometimes be masked by adding a large excess of one of the reaction products. When Fe(II), Sn(II), or Sb(III) are titrated with ceric sulfate, any arsenite in the solution interferes because of the following reaction:

$$AsO_2^- + 2\,Ce^{4+} + 2\,H_2O \rightarrow AsO_4^{3-} + 2\,Ce^{3+} + 4\,H^+$$

Addition of sufficient AsO_4^{3-} to the system displaces the oxidation–reduction potential so that this reaction no longer competes with the principal reaction.

Such changes in oxidation–reduction potentials are exploited in chemical analysis. In the presence of EDTA, which also serves to mask Bi, Cd, Cu, and Pb, ferrous sulfate becomes a much more powerful reducing agent, depositing Ag as a black precipitate from Ag(I) solutions in sodium acetate, and making possible in this way the potentiometric or amperometric titration of Ag.[136] In the titrimetric determination of glucose and other reducing sugars by Bertrand's method, Cu tartrate can be replaced by CuEDTA or CuNTA.[137,138]

The photometric titration of Co(II) with hexacyanoferrate is based on the following reaction:

$$Fe(CN)_6^{3-} + Co^{2+} \rightarrow Co^{3+} + Fe(CN)_6^{4-}$$

Interference by Mn(II) can be avoided by masking with glycine at pH 8.[139]

Phosphate or fluoride ions mask Fe(III) in the precipitation titration of chloride ion with Hg(I), against a chelatochrome indicator.[140] Fluoride ion is also a masking agent in a volumetric method for Al in mineral ores. When fluoride ion is added to a solution containing aluminate ions, aluminum–fluoride complexes are formed with the liberation of hydroxyl ions which can be titrated.[141] Alternatively, Al can be determined volumetrically on the basis of a reaction of Al(III) with alkaline citrate to form the citrate complex and generate hydroxyl ions.[142]

The preferential masking of Al by fluoride ion in a mixture of Al and Fe(III) provides a potentiometric method for the determination of Al by titration with fluoride ion in the presence of a little Fe(II). The Fe(III)/Fe(II) potential remains constant until all of the Al is masked by fluoride ion, then changes rapidly because Fe(III) fluoride complexes are formed.[143] The same reaction has been made the basis of an amperometric method.[144] A similar principle applies to the amperometric titration of calcium with EDTA in the presence of zinc as indicator, the latter being partially masked by ammonia or hydroxide ions.[145,146]

Free ammonia in sodium citrate solutions cannot be titrated directly because of buffering by citrate ion, but this difficulty can be overcome by adding excess calcium chloride to mask citrate as its stable calcium complex.[147] The Wollack titration of thiosulfate is a further example of the masking of anions or neutral molecules. In this case, free iodine is masked by adding sodium sulfite, and any free sulfite is then masked by complexing with formaldehyde.[148]

Recently, a method has been described for the consecutive titration of iodate and periodate with thiosulfate ion.[149] At pH 2.9, molybdate in excess masks periodate against reaction with iodide ion, by forming 6-molybdo periodate, $[I(MoO_4)_6]^{5-}$, but does not interfere in the reaction between iodate and iodide ions. Addition of oxalic acid then demasks the periodate which, in turn, reacts with iodide ion.[149] The same method has been used to detect iodate formed in the reaction of periodate with *vic*-dihydroxy compounds.[150] Tungstate can replace molybdate, but less effectively.

REFERENCES

1. T. A. Kiss, *Z. Anal. Chem.* **208,** 334 (1965).
2. I. Sajo, *Magyar Kem. Folyoirat* **60,** 268, 331 (1954); **62,** 56 (1956).
3. T. N. Nazarchuk, and L. N. Mekhanoshina, *Zh. Analit. Khim.* **20,** 262 (1965).
4. R. Pribil, and V. Vesely, *Chemist-Analyst* **55,** 38 (1966).
5. R. Pribil, and V. Vesely, *Talanta* **9,** 23 (1962).
6. Ref. 4, p. 68.
7. R. Pribil, and V. Vesely, *Talanta* **10,** 1287 (1963).
8. D. H. Wilkins, *Anal. Chim. Acta* **23,** 309 (1960).
9. S. M. Bresler, and V. A. Rogozhina, *Nauchno-issled. Trudy tsentr. nauchno-issled. Inst. kozk.-obuv. Prom.* **1966,** 44.
10. I. M. Yurist, *Zavodskaya Lab.* **32,** 1050 (1966).

11. V. N. Alferova, T. F. Davidenko, and L. I. Serikova, *Khim. Prom. Ukrainy, nauchno-proiz. Sb.* **5,** 46 (1966).
12. L. Szekeres, E. Kardos, and G. L. Szekeres, *Chemist-Analyst* **53,** 72 (1964).
13. E. A. Shteiman, Z. G. Dobrynina, and E. A. Mordovskaya, *Zavodskaya Lab.* **30,** 1200 (1964).
14. L. Barcza, and E. Körös, *Chemist-Analyst* **48,** 94 (1959).
15. R. Pribil, and V. Vesely, *Chemist-Analyst* **54,** 12 (1965).
16. G. W. C. Milner, and J. W. Edwards, *Anal. Chim. Acta* **18,** 513 (1958).
17. A. Bacon, and G. W. C. Milner, *A.E.R.E. Report C/R 1992* (1956).
18. R. Pribil, *Coll. Czech. Chem. Comm.* **18,** 783 (1953).
19. R. Fabregas, A. Prieto, and C. Garcia, *Chemist-Analyst* **51,** 77 (1962).
20. R. Pribil, and V. Vesely, *Talanta* **12,** 475 (1965).
21. T. Mekada, K. Yamaguchi, and K. Ueno, *Talanta* **11,** 1464 (1964).
22. H. Flaschka, and J. Butcher, *Microchem. J.* **7,** 407 (1963); *Talanta* **11,** 1071 (1964).
23. Ref. 4, p. 4.
24. G. Böltz, H. Wiedmann, and W. Kurela, *Metall* **10,** 821 (1956).
25. R. Belcher, R. A. Close, and T. S. West, *Talanta* **1,** 238 (1958).
26. L. A. Vol'f, *Zavodskaya Lab.* **25,** 1438 (1959); Y. V. Morachevsky, and L. A. Vol'f, *Zh. Analit. Khim.* **15,** 656 (1960).
27. K. Yamaguchi, and K. Ueno, *Talanta* **10,** 1195 (1963).
28. H. L. Matts, *Anal. Chem.* **32,** 1189 (1960).
29. G. Brunisholtz, M. Genton, and E. Plattner, *Helv. Chim. Acta* **36,** 782 (1953).
30. R. Pribil, *Coll. Czech. Chem. Comm.* **16,** 86 (1951).
31. R. A. Burg, and H. F. Conaghan, *Chemist-Analyst* **49,** 100 (1960).
32. A. Ringbom, G. Pensar, and E. Wanninen, *Anal. Chim. Acta* **19,** 525 (1956).
33. R. Pribil, and V. Vesely, *Talanta* **13,** 233 (1966).
34. L. Szekeres, E. Kardos, and G. L. Szekeres, *Chemist-Analyst* **54,** 53 (1965).
35. A. M. Escarrilla, *Talanta* **13,** 363 (1966).
36. T. Mekada, G. Yamaguchi, and K. Ueno, *Talanta* **11,** 1461 (1964).
37. J. Kinnunen, and B. Wennestrand, *Chemist-Analyst* **46,** 92 (1957).
38. R. Pribil, and V. Vesely, *Chemist-Analyst* **50,** 100 (1961).
39. M. Pospisil, and J. Dolezal, *Chem. Listy* **59,** 1472 (1965).
40. R. Pribil, and V. Vesely, *Talanta* **8,** 565 (1961).
41. R. Pribil, *Coll. Czech. Chem. Comm.* **19,** 1171 (1954).
42. E. Lassner, and R. Scharf, *Chemist-Analyst* **49,** 44 (1960).
43. A. I. Busev, V. M. Ivanov, and L. L. Talipova, *Zh. Analit. Khim.* **18,** 33 (1963).
44. R. Pribil, and V. Vesely, *Talanta* **8,** 880 (1961); *Chemist-Analyst* **53,** 38 (1964); *Talanta* **9,** 1053 (1962).
45. A. A. Ashton, *Anal. Chim. Acta* **28,** 296 (1963).
46. C. Liteanu, C. Murgu, and L. Marinescu, *Z. Anal. Chem.* **175,** 1 (1960).
47. J. Janousek, and K. Studlar, *Coll. Czech. Chem. Comm.* **24,** 3799 (1959).
48. R. Pribil, and V. Vesely, *Chemist-Analyst* **52,** 5 (1963).
49. R. Pribil, and V. Vesely, *Chemist-Analyst* **53,** 38 (1964).

50. E. Lassner, and R. Scharf, *Chemist-Analyst* **50,** 5 (1961).
51. J. Dolezal, V. Patrovsky, Z. Sulcek, and J. Svasta, *Coll. Czech. Chem. Comm.* **21,** 979 (1956).
52. J. E. Mee, and J. D. Corbett, *Chemist-Analyst* **50,** 74 (1961).
53. H. Flaschka, and J. Garrett, *Z. Anal. Chem.* **218,** 338 (1966).
54. K. Mizuno, *Bunseki Kagaku* **14,** 410 (1965).
55. K. L. Cheng, and B. L. Goydish, *Talanta* **13,** 1161 (1966).
56. J. Kinnunen, and B. Merikanto, *Chemist-Analyst* **44,** 11 (1955).
58. D. G. Biechler, *Chemist-Analyst* **52,** 48 (1963).
59. H. Khalifa, and M. M. Khater, *Z. Anal. Chem.* **184,** 92 (1961).
60. R. S. Volodarskaya, N. A. Kanaev, and G. N. Derevyanko, *Zavodskaya Lab.* **32,** 413 (1966).
61. H. Flaschka, and A. M. Amin, *Z. Anal. Chem.* **140,** 6 (1953).
62. Ref. 14, p. 69.
63. Ref. 15, p. 46.
64. R. Pribil, and V. Vesely, *Talanta* **8,** 743 (1961); **9,** 23 (1962).
65. H. R. Das, and S. C. Shome, *Anal. Chim. Acta* **35,** 256 (1966).
66. H. Flaschka, *Mikrochim. Acta* **1955,** 55.
67. S. J. Lyle, and M. M. Rahman, *Talanta* **10,** 1183 (1963).
68. R. Pribil, and V. Vydra, *Coll. Czech. Chem. Comm.* **24,** 3103 (1959).
69. A. F. Kuteinikov, and V. M. Brodskaya, *Zavodskaya Lab.* **28,** 792 (1962).
69a. R. Pribil, *Talanta* **14,** 619 (1967).
70. Y. V. Morachevskii, and L. A. Vol'f, *Isv. vyssh. ucheb. Zaved., Khim. khim. Tekhnol.* **7,** 513 (1964).
71. G. W. C. Milner, and J. W. Edwards, *Anal. Chim. Acta* **18,** 513 (1958).
72. H. Shui-Chieh, and L. Shu-Chuan, *Scientia Sinica* **13,** 1619 (1964).
73. R. Pribil, and V. Vesely, *Talanta* **9,** 939 (1962); **10,** 899 (1963).
74. G. S. Tereshin, and I. V. Tananaev, *Zh. Analit. Khim.* **17,** 526 (1962).
75. O. I. Milner, and S. J. Gedansky, *Anal. Chem.* **37,** 931 (1965).
76. A. Brück, and K. F. Lauer, *Anal. Chim. Acta* **33,** 338 (1965).
77. H. Flaschka, and F. Huditz, *Z. Anal. Chem.* **137,** 172 (1952).
78. C. Harzdorf, *Z. Anal. Chem.* **203,** 101 (1964).
78a. J. C. Cornwell, and K. L. Cheng, *Anal. Chim. Acta* **42,** 189 (1968).
79. K. E. Burke, *Anal. Chim. Acta* **34,** 485 (1966).
80. H. Flaschka, *Chemist Analyst* **42,** 56 (1953).
81. E. Körös, *Ann. Univ. Sci. Budapest, Sect. Chim.* **2,** 269 (1960).
82. I. M. Yurist, *Zh. Analit. Khim.* **22,** 442 (1967).
83. R. Pribil, and M. Kopanica, *Chemist-Analyst* **48,** 35 (1959).
84. H. Flaschka, and R. Püschel, *Chemist-Analyst* **44,** 71 (1955).
85. R. Pribil, and V. Vesely, *Talanta* **12,** 385 (1965).
86. P. Povondra, and R. Pribil, *Coll. Czech. Chem. Comm.* **26,** 2164 (1961).
87. R. Pribil, *Talanta* **13,** 1223 (1966).
88. J. Kinnunen, and B. Merikanto, *Chemist-Analyst* **43,** 93 (1954).
89. K. Ueno, *Anal. Chem.* **29,** 1668 (1957).
90. F. Sierra, and G. Asensi, *Anales Real Soc. Espan. Fis. Quim. (Madrid), Ser. B.* **57,** 435 (1961).

91. Ref. 33, p. 515.
92. E. S. Bruile, and K. S. Merkulova, *Zh. Prikl. Khim.* **37,** 216 (1964).
93. C. A. Goetz, and F. J. Debbrecht, *Anal. Chem.* **27,** 1972 (1955).
94. Ref. 40, p. 880.
95. D. H. Wilkins, *Anal. Chim. Acta* **23,** 309 (1960).
96. V. Patrovsky, *Z. Anal. Chem.* **214,** 261 (1965).
97. H. Flaschka, and F. Sadek, *Chemist-Analyst* **47,** 30 (1958).
98. A. Holasek, H. Lieb, and M. Pecar, *Mikrochim. Acta* **1960,** 750.
99. R. Pribil, and V. Vesely, *Talanta* **11,** 1545 (1964).
100. Ref. 15, p. 100.
101. R. Pribil, and J. Horacek, *Talanta* **14,** 313 (1967).
102. R. Pribil, V. Vesely, and J. Horacek, *Talanta* **14,** 266 (1967).
103. H. Flaschka, *Mikrochim. Acta* **1952,** 391.
104. M. A. Wade, and H. J. Sein, *Anal. Chem.* **33,** 793 (1961).
105. F. W. E. Strelow, and F. von S. Toerien, *Anal. Chim. Acta* **36,** 189 (1956).
106. K. L. Cheng, *Microchem. J.* **8,** 225 (1964).
107. G. W. C. Milner, and J. W. Edwards, *Anal. Chim. Acta* **20,** 31 (1959).
108. G. W. C. Milner, and G. A. Barnett, *A.E.R.E. C/R 1865* (1956); E. Lassner, and R. Scharf, *Chemist-Analyst* **50,** 6 (1961).
109. A. C. S. Costa, and T. Mascarenhas, *An. Ass. Brasil Quim.* **26**(3,4), 35 (1967).
110. I. Dubsky, *Coll. Czech. Chem. Comm.* **24,** 4045 (1959).
111. L. J. Ottendorfer, *Chemist-Analyst* **47,** 96 (1958).
112. B. Bieber, and Z. Vecera, *Coll. Czech. Chem. Comm.* **26,** 2031 (1961).
113. Y. C. Chen, and H. J. Li, *Hua Hsueh Hsueh Po* **31,** 391 (1965).
114. E. S. Bruile, *Zh. Prikl. Khim., Leningrad* **39,** 1192 (1966).
115. B. Budesinsky, A. Bezdekova, and D. Vrzalova, *Coll. Czech. Chem. Comm.* **27,** 1528 (1962).
116. B. H. J. de Heer, T. van der Plas, and M. E. A. Hermans, *Anal. Chim. Acta* **32,** 292 (1965).
117. D. Filipov, and N. Kirtcheva, *Compt. Rend. Acad. Bulg. Sci.* **17,** 467 (1964).
118. M. P. Borzenkova, *Vest. Mosk. Gos. Univ., Ser. Khim.* **1966,** 50.
119. J. J. Hickey, and C. J. Overbeck, *Anal. Chem.* **38,** 932 (1966).
119a. T. A. Kiss, L. J. Bjelica, and E. S. Kervesan, *Mikrochim. Acta,* **1968,** 1284.
120. R. Pribil, *Coll. Czech. Chem. Comm.* **20,** 162 (1955).
121. A. A. Ashton, *Anal. Chim. Acta* **28,** 296 (1963).
122. S. Hara, *Bunseki Kagaku* **10,** 633 (1961).
123. V. Riha, S. Kotrly, and J. Vesely, *Mikrochim. Acta* **1966,** 342.
124. T. Mekada, K. Yamaguchi, and K. Ueno, *Talanta* **11,** 1461 (1964).
125. L. E. Pechentkovskaya, and T. N. Nazarchuk, *Zh. Analit. Khim.* **19,** 897 (1964).
126. G. W. C. Milner, and J. W. Edwards, *Anal. Chim. Acta* **20,** 31 (1959).
127. T. Miyajima, *Bunseki Kagaku* **12,** 742 (1963).
128. R. M. Dagnall, T. El-Ghamry, and T. S. West, *Talanta* **13,** 1667 (1966).
129. H. Khalifa, and B. Ateya, *Microchem. J.* **12,** 440 (1967).
130. L. E. Kalbus, and G. E. Kalbus, *Anal. Chim. Acta* **39,** 335 (1967).

131. J. Havir, *Coll. Czech. Chem. Comm.* **32,** 130 (1967).
132. E. Bakacs-Polgar, L. Szekeres, and B. Lang, *Z. Anal. Chem.* **158,** 14 (1957).
133. L. Szekeres, E. Kardos, and G. L. Szekeres, *Chemist-Analyst* **53,** 40 (1964).
134. R. Pribil, and L. Wünsch, *Chem. Listy* **46,** 337 (1952); *Coll. Czech. Chem. Comm.* **18,** 306 (1953).
135. D. D. Perrin, *Organic Complexing Reagents,* Interscience, New York, 1964, Chap. 4.
136. R. Pribil, J. Dolezal, and V. Simon, *Chem. Listy* **47,** 1017 (1953).
137. H. Yoshida, S. Murakami, and H. Hayakawa, *Kobe Daigaku Kyoikugakubu Kenkyu Shuroku Shizen Kagakuhen* **33,** 137 (1965).
138. C. Hennart, *Talanta* **12,** 420 (1965).
139. H. Poppe, and G. den Boef, *Talanta* **12,** 625 (1965).
140. R. Püschel, and E. Lassner, *Chemist-Analyst* **50,** 26 (1961).
141. H. L. Watts, *Anal. Chem.* **30,** 967 (1958).
142. C. L. Wilson, and D. W. Wilson, *Comprehensive Analytical Chemistry,* American Elsevier Publishing Co., New York, 1962, Vol. 1c, p. 103.
143. W. D. Treadwell, and E. Bernasconi, *Helv. Chim. Acta* **13,** 500 (1930).
144. A. Ringbom, and B. Wilkman, *Acta Chem. Scand.* **3,** 22 (1949).
145. R. Pribil, and E. Vincenova, *Coll. Czech. Chem. Comm.* **18,** 308 (1953).
146. H. A. Laitinen, and R. F. Sympson, *Anal. Chem.* **26,** 556 (1954).
147. A. Ringbom, *Pure and Appl. Chem.* **7,** 473 (1963).
148. G. W. Latimer, *Talanta* **13,** 321 (1966).
149. R. Belcher, and A. Townshend, *Anal. Chim. Acta* **41,** 395 (1968).
150. G. Nisli, and A. Townshend, *Talanta* **15,** 1377 (1968).

CHAPTER

6

MASKING IN PRECIPITATION REACTIONS

A species ML_n cannot be precipitated from solution unless the quantity $[M][L]^n$ exceeds the solubility product, K_{sp}, where the square brackets denote concentrations of free metal ion, M, and an anion, L. By adding suitable complexing agents it is often possible to diminish the concentration of one of these reactants sufficiently so that precipitation is prevented or even to secure the dissolution of a precipitate that has already been formed. In a limited number of cases this effect can be achieved by adding a sufficient excess of the complexing agent that is also being used as a precipitant. Thus, silver cyanide is only sparingly soluble in water, but if excess cyanide ion is added, silver cyanide redissolves to form $Ag(CN)_2^-$ in which the silver is effectively masked against precipitation.

Usually, however, the masking agent is a different species, and its most common applications in analytical chemistry are for the avoidance of hydrolytic precipitation and for the achievement of adequate selectivity. The problem of finding a suitable reagent to mask metal[1] while metal[2] is precipitated requires that the complexes with metal[1] should be both strong and soluble, whereas the complexes with metal [2] should be weak enough so that the precipitation reaction is not impeded. Solubility of the metal[1] complex can be achieved by ensuring that the complex bears a net electronic charge or else that the ligand has many polar groups to ensure strong hydrogen bonding to solvent water. Qualitative considerations discussed in Chapters 2 and 3 suggest ligands that are likely to be successful, such as polyamines with Cu(II) or di- and tricarboxylic acids with tervalent cations.

Thus Fe(III) ion forms a much more stable, water-soluble complex with tartaric acid than Ni(II) does, and it can be masked in this way when Ni(II) is precipitated as its dimethylglyoxime complex. Similarly, Fe(III), Al(III), and Mn(III) form strong complexes in alkaline solution with triethanolamine, whereas the Mg(II) complex is weak. This enables Mg(II) to be precipitated from ammoniacal solution as $MgNH_4PO_4$ while the other metal ions remain in solution.[1]

Changes in valence state are sometimes important for successful masking. Reducing agents such as hydrazine, hydroxylamine hydrochloride, or ascorbic acid reduce Fe(III) to Fe(II), or prevent the oxidation of Mn(II) to Mn(III). Cyanide ion reduces Cu(II), forming soluble Cu(I) complexes and masking Cu(I) against precipitation as Cu_2S (which is appreciably more soluble than CuS). The Cd(II) cyanide complexes, on the other hand, are not sufficiently stable to prevent the precipitation of CdS, so that cyanide ion is a suitable masking agent for Cu(II) when Cd is precipitated in this way. As an example of an oxidizing agent, in acid solution hydrogen peroxide converts molybdate to permolybdic acid, thereby masking it against precipitation by Ba(II), Pb(II), benzoinoxime, or methylene blue, or against phosphomolybdate formation in the test for phosphate. Similarly, when Zr(IV) is precipitated as the hydroxide from ammoniacal solution, Nb and Ta can be retained in solution by the addition of hydrogen peroxide to convert them to soluble peroxy complexes.[2]

The quantitative treatment of masking in precipitation reactions is straightforward, given a knowledge of stability constants and solubility products. The precipitation of barium iodate from a solution containing barium and calcium ions at concentrations of about $0.01M$ can be taken as an example. The solubility products of $Ba(IO_3)_2$ and $Ca(IO_3)_2$ are 6×10^{-10} and 6×10^{-7}, respectively, so that in the presence of M iodate the concentrations of free barium and calcium ions would also be 6×10^{-10} and $6 \times 10^{-7}M$. Both of these cations would be quantitatively precipitated. By prior addition of EDTA in $0.1M$ excess and adjustment of the pH of the solution to 7, the free calcium ion would be diminished to $3 \times 10^{-9}M$ but the free barium ion would be $2 \times 10^{-6}M$. The calcium level is now too low by a factor of 200 for calcium iodate to precipitate, whereas the barium ion is present at more than 3000 times its equilibrium concentration, so that it would be precipitated quantitatively.

The effectiveness of a masking agent is almost always pH-dependent because, with few exceptions, pK values for proton addition to ligands occur at or near the pH range over which they are ordinarily used for complex formation. The precipitant is frequently the anion of a weak acid, so that the solubility of the complex is also pH-dependent. To take account of these factors, the concept of conditional constants is useful. Examples are available describing the calculation, for particular cases, of the effectiveness of masking agents, or of methods based on them.[3]

MASKING AGAINST HYDROLYTIC PRECIPITATION

The sparing solubility of many metal hydroxides sets an upper limit on the permissible concentration of free metal ions in a solution if precipitation is to be avoided. For a hydroxide $M(OH)_n$ the relation is

$$\log [M^{n+}]_{max} = n \cdot pK_w - pK_{sp} - n \cdot pH$$

so that for a species such as $Fe(OH)_3$ an increase of 1 in the pH of the solution decreases the maximum permissible value of $[Fe^{3+}]$ by a factor of 1000. This equation is an oversimplification in cases where polynuclear complexes and other hydrolyzed species are also formed, but it is adequate under conditions where the free metal ion concentration is kept low, such as by the addition of a complexing agent.

From the approximate values of pK_{sp} given in Table 6.1 the absence of complexing agents limits the ordinary working conditions for higher

Table 6.1. Solubility Products of Metal Hydroxides at 20–25°[a]

Hydroxide	pK_{sp}	Hydroxide	pK_{sp}
AgOH	7.6	$Mg(OH)_2$	10.4
$Al(OH)_3$	31.6	$Mn(OH)_2$	12.3
$Au(OH)_3$	45.6	$Nd(OH)_3$	21.5
$Be(OH)_2$	21.2	$Ni(OH)_2$	16.8
BiO(OH)	9.4	$Pb(OH)_2$	15.7
$Ca(OH)_2$	4.9	$Pb(OH)_4$	65.5
$Cd(OH)_2$	13.2	$Pd(OH)_2$	6.8
$Ce(OH)_3$	19.5	$Pr(OH)_3$	21.2
$Ce(OH)_4$	50.4	$Pt(OH)_2$	35
$Co(OH)_2$	15.3	$Pu(OH)_2$	19.7
$Co(OH)_3$	43.8	$Ru(OH)_3$	36
$Cr(OH)_2$	16.6	$Sb(OH)_3$	4.7
$Cr(OH)_3$	30.3	$Sc(OH)_3$	29.4
$Cu(OH)_2$	19.7	$Sn(OH)_2$	27.7
$Er(OH)_3$	23.4	$Sn(OH)_4$	56
$Eu(OH)_3$	23.1	$Th(OH)_4$	44
$Fe(OH)_2$	14.7	$TiO(OH)_2$	28.6
$Fe(OH)_3$	37.9	$Tl(OH)_3$	45.2
$Ga(OH)_3$	35.7	$Tm(OH)_3$	23.5
$Gd(OH)_3$	22.7	$UO_2(OH)_2$	21.6
$Hf(OH)_4$	25.4	$VO(OH)_2$	22.1
$Hg_2(OH)_2$	46[b]	$Y(OH)_3$	22.8
$Hg(OH)_2$	25.0	$Yb(OH)_3$	23.6
$In(OH)_3$	32.9	$Zn(OH)_2$	15.6
$La(OH)_3$	19	$ZrO(OH)_2$	47
$Lu(OH)_3$	23.7		

[a] "Mixed" constants, usually at $\mu = 0.1$ and for aged precipitates.
[b] For $Hg_2(OH)_2 \rightleftharpoons Hg_2^{2+} + 2OH^-$.

valent and transition metal ions to acid or weakly basic solutions. Precipitation can be avoided, and the accessible pH range extended, by the addition of suitable masking agents. This also diminishes the formation of basic salts such as BiOCl and partially hydrolyzed complexes. In the few cases where the hydroxide is amphoteric (eg, $Al(OH)_3$, $Zn(OH)_2$), it can be dissolved by raising the pH further to form anionic complexes such as aluminates and zincates.

Tartaric, citric, and catecholdisulfonic acids are effective masking agents for maintaining many metal ions in solution, and this has been brought out in studies with Bi(III), ZrO^{2+}, Ga(III), Tl(III), La(III), and Ce(IV). For Bi(III) the sequence was as follows:[4]

catecholdisulfonic acid > tartaric acid or cysteine > citric acid > EDTA > dimercaptopropanol > sulfosalicylic acid > sodium thiosulfate > potassium iodide > potassium thiocyanate

With ZrO^{2+} masking agents formed the following sequence:[4]

catecholdisulfonic acid > citric acid > tartaric acid > EDTA > malic acid > sulfosalicylic acid > sodium carbonate > sodium fluoride > catechol or inositol > sodium salicylate or cysteine > oxalic acid

Other sequences were,[5] for Al(III),

citric acid > tartaric acid > sulfosalicylic acid > EDTA > sodium salicylate

for Ga(III),

citric acid > EDTA > sulfosalicylic acid = tartaric acid > oxalic acid

and for Tl(III),

EDTA > oxalic acid > tartaric acid

Similarly, catecholdisulfonic acid, citric acid, and EDTA mask Pb(II), Co(II), Mg(II), Cr(III), and Be(II) against hydrolytic precipitation.[6] Tartaric, citric and catecholdisulfonic acids have a high capacity for masking Mn(II) and Cu(II) in this way, but the colorless Mn(II) complexes are readily oxidized by air to give colored Mn(III) complexes unless hydroxylamine or hydrazine is also present.[7] Catecholdisulfonic acid and EDTA are useful for preventing the precipitation of La(III) and Ce(IV), but oxalic, phosphoric, and salicylic acids are much less effective.[8]

Triethanolamine masks Fe(III) against precipitation by sodium hydroxide when Ti(IV)[9] and La(III)[10] are deposited as their hydroxides. Divalent metal ions and Al(III) are also masked.[10] Glycine can be used to decrease the coprecipitation of Cu(II), Zn(II), Cd(II), Ni(II), and Co(II) when Al(III) is precipitated as its hydroxide.[11] Potassium trithiocarbonate is a suitable masking agent for Fe(III) when Ti(IV),

Be(II), and Zr(IV) are precipitated as their hydroxides.[12] Conversely, potassium trithiocarbonate can be used as a precipitant for Tl(I) in slightly ammoniacal solution, if potassium cyanide is added to mask Ag, Hg(II), Cu(II), Fe(III), Co(II) and Ni, and tartaric acid is present to mask Al, Cr(III), Be, Ti, and Fe(III). EDTA masks Fe(III), Co(II), Ni, Zn, Ba, Ca, Sr, and Mg.[12a]

A more general masking agent is EDTA, which can frequently be applied in cases where the metal ion to be precipitated forms only a comparatively weak EDTA complex. The difference in stability constants of the Ca and Mg EDTA complexes enables Ca(II) to be masked by a rough titration with EDTA (against murexide) without preventing the subsequent precipitation of $Mg(OH)_2$.[1] Similarly, because their EDTA complexes are relatively weak, Be(II), Ag(I), Nb(V), Ta(V), and Hg(II) can be precipitated as their hydroxides from neutral or slightly acid EDTA solutions.[13] Use of EDTA reduced the amount of coprecipitation of Bi(III), Pb(II), and Cd(II) when $Sn(OH)_4$ was precipitated from ammoniacal solution,[14] and also retained In(III) in solution.[15] Separation of Ga(III) from Fe(III) can be achieved by masking Ga with EDTA before precipitating $Fe(OH)_3$ with sodium hydroxide at pH 12.5.[16] Likewise, EDTA masks Mo(VI) and Al(III) when Nb(V) is precipitated as its hydroxide at pH 10 by adding ammonia.[17]

EDTA AS A MASKING AGENT IN PRECIPITATIONS OF METAL COMPLEXES

QUALITATIVE APPLICATIONS[18,19]

Increased selectivity is often possible when precipitations are carried out in the presence of EDTA as a masking agent.[20] For example, chloride ion precipitates Ag(I), Tl(I), Hg(I), and Pb(II) from acid solution, but if EDTA is present Hg(I) and Pb(II) are masked and remain in solution. Similarly, in acid solution Ag(I) and Tl(I) are precipitated by iodide ion, whereas Bi(III) and Pb(II) are masked by EDTA. EDTA is also suitable for masking Tl(I) in ammoniacal solution when Ag(I) is detected by precipitation with iodide ion. Precipitation of Tl(I) and Ba(II) as their chromates takes place from acetic acid solution in the presence of EDTA which retains Ag(I) and Pb(II) in the solution.

Addition of EDTA leads to masking of Al(III), Bi(III), Cr(III), Fe(III), Hg(II), and Mn(III) when the hydroxides of Be(II), Sb(III), Sn(IV), Ti(IV), and U(VI) are precipitated from ammoniacal solution. Zinc(II) and Fe(III) sulfides are precipitated from sodium sulfide solution while, in the presence of EDTA, Ni(II), Co(II) and Mn(II) are masked. If excess calcium chloride is then added, Co(II) and Mn(II), but not Ni(II), are demasked and can also be precipitated. Similarly, EDTA masks Zn against precipitation by hydrogen sulfide or sodium diethyldithiocarbamate in ammoniacal solutions, but addition of Ca displaces the Zn, which then reacts normally.

GRAVIMETRIC METHODS

Many gravimetric methods use EDTA to improve their selectivity. The quantitative treatment of equilibria in such systems has been developed by Hulanicki,[21] Ringbom,[3] and Kelly and Sutton.[22] The latter have calculated the concentrations of metals that should remain in solution at different pH values for systems such as Ba–Ca–Sr in the presence of EDTA and sulfate ion; Co–Cu–Ni with EDTA and 8-hydroxyquinoline; Co–Cd–Zn with EDTA and sulfide ion; Ce–La–Y with EDTA and oxalate ion; and Ba–Ca–Sr with EDTA and oxalate ion.[22] Ryan[22a] has derived a more general form of Hulanicki's masking coefficient equation[21] which is useful for calculating the amount of chelating agent necessary for complete masking in precipitation reactions, so long as it can be assumed that the only important side reaction of the metal ion is the formation of a 1:1 complex. This condition is probably met, for example, when ligands such as EDTA are used as masking agents.

Cheng[13] has discussed examples in which EDTA can be used as a masking agent to increase selectivity in the precipitation of metal ions by halides, sulfate, and tellurite ions. Controlling factors are the pH at which the precipitation is carried out, and the concentrations of EDTA and the precipitant. The fluorides of Th(IV), La(III), Sc(III), Y(III) and the rare earths can be separated from Mg(II) and Ca(II) when EDTA is used as a masking agent, and barium sulfate is precipitated in the presence of Sr and Pb.[13]

Beryllium can be determined in the presence of metals such as Al, Ca, Cu, Fe, Mg, Ni, and Zn by precipitation as the hydroxide or with a saturated hexammine cobalt(III) solution, if EDTA is used as a masking agent.[23,24,25] The final estimation is gravimetric or, if the cobaltammine is used, by complexometric titration of the Co in the precipitate.[25,26] Possible interference by EDTA was avoided in a gravimetric method for Be, based on precipitation with benzoylacetanilide at pH 7.0–7.9. The method used MgEDTA as a masking agent because its stability constant is greater than for BeEDTA, but the Mg can be displaced by the metals to be masked, namely, Cu, Ni, Co, Zn, Al, Fe, Cr, Ce, V(IV), Mo(VI), and W(VI).[27] Other masking agents included tartrate ion for Cu, Co, Zn, and U(VI), tartrate and cyanide for higher levels of Fe(III), and MgEDTA with hydrogen peroxide, or 7-iodo-8-hydroxyquinoline-5-sulfonic acid (ferron), for Ti(IV).[27] Earlier examples of masking with EDTA in the gravimetric determination of Be, and in the separation of Be, U, and Ti, are given in references 28–33.

EDTA can be used as a general masking agent when Ca is precipitated as the oxalate.[34] Conversely, Ca can be masked by EDTA against coprecipitation in the separation of Ba as chromate, Sr as sulfate, or Mg as hydroxide.[35] Barium is also masked by EDTA when Ra is precipitated as its sulfate.[36]

8-Hydroxyquinoline is not selective as a precipitant for Al(III), so that careful masking is necessary. The reaction is carried out in an ammonium citrate buffer of pH 8.5–10, containing added EDTA, cyanide ion, and sodium sulfite. EDTA masks Cd, Mn, Pb, Zn, rare earths, alkaline earths, and Mg; citrate also masks Fe and rare earths, while cyanide masks Cu, Ni, and Fe, the sulfite reducing $Fe(CN)_6^{3-}$ to $Fe(CN)_6^{4-}$.[37] If Cr(III) is present it can be masked by oxidation to chromate or by boiling with EDTA to form a kinetically inert complex.

At pH 5.5, cupferron forms insoluble precipitates with Ti, Zr, Hf, Nb, Ta, and many other metal ions. Selectivity is improved by adding EDTA, fluoride ion (which masks Be, Al, Ce(III), U(VI), and Sn(IV), and citric acid.[38]

Masking with EDTA or reduction with ascorbic acid overcomes interference by Fe(III) in the determination of W(VI) by precipitation at pH 5 with variamine blue.[39] A gravimetric method for U(VI) with 4-aminobenzoic acid at pH 3–5 uses EDTA dissolved in aqueous pyridine as a masking agent.[40] EDTA was also present when U(VI) was precipitated as ammonium diuranate with ammonia.[41]

Numerous reagents have been proposed for the precipitation of Pd(II), and in many cases, EDTA has been used to mask Fe, Cu, and other cations. These reagents have included quinolinic acid,[42] α-picolinic acid (pH 3–7),[43] β-aminopicolinic acid (pH above 3),[44] 2-hydroxy-5-methylpropiophenone oxime (pH 3),[45] di-2-pyridyl ketoxime (pH 4–11)[46], and diethyldithiophosphoric acid (pH 9).[47] Tartrate[43,47] and citrate[46] were also included as masking agents and as buffers.

Masking with EDTA and tartrate at pH 8.6–10 makes benzimidazol-2-ylmethanethiol a highly selective precipitant for Ag(I).[48] Interference by Hg(I) can be overcome by oxidation to Hg(II).[48] Mercaptophenylthiodiazolone has also been proposed as a precipitant for Ag(I)[49] and Hg(I),[50] with EDTA masking.

EDTA and nitrilotriacetic acid prevent the precipitation of basic Bi salts, and also mask interfering ions, in the precipitation of Bi(III) with dimethylglyoxime.[51] Propylenediaminetetraacetic acid has been used instead of EDTA, with the inclusion of cyanide ion to mask Ni, Co, Cu, Pd, Zn, Ag, Cd, and Hg, and dimethylglyoxime has been replaced by oximes derived from large ring 1,2-diketones.[52] Cyclohexanediaminetetraacetic acid was too strong a complexing agent to be used for masking, and hydroxyethylethylenediaminetriacetic acid interfered in the determination. In the gravimetric determination of Bi with N-benzoyl-*N*-phenylhydroxylamine at pH 6.0–6.8, EDTA masked Bi as well as other metals. When Ca(II) was added, Bi was demasked and precipitated.[53] Other masking agents that were compatible included tartrate and citrate (for Be, Pb, Sb, La, Th, As, U) or tartrate and oxalate (for Sn(IV) and Al). Thioglycolic acid could not be used because it masked Bi too strongly.[53]

The selectivity of the piazselenol method for Se, using 4,5-dichloro-*o*-phenylenediamine, is improved by masking Zn, Cd, Cu, Mg, Ca, In, and other cations with EDTA.[54] Similarly, a mixture of EDTA, sodium fluoride, and sodium oxalate was added when Se was determined gravimetrically, spectrophotometrically, or fluorometrically with 2,3-diaminonaphthalene.[55]

OTHER MASKING REAGENTS

Tartaric acid is a versatile masking agent for higher-valent ions. For example, in acid solution (pH 1.0–1.5) it masks Fe(III) when Ti(IV) is precipitated with benzoylacetanilide.[27] At pH 6.5, Al(III) is masked during the precipitation of Zn(II) with 8-hydroxyquinaldine,[56] and at pH 12–13 tartrate masks As(III), Sb(III), Sn(IV), Bi(III), and Zn(II) when Ni(II) is precipitated with *N,N'*-di(salicylidene)ethylenediamine.[57] In the latter case further masking is achieved with fluoride ion (for Sb(III) and Sn(IV)), triethanolamine (for Fe(III)), and potassium iodide (for Hg(II)).[57] *o*-Hydroxyacetophenone oxime has been used to determine Pd(II), with tartaric acid to mask Fe(III), Sb(III), Bi(III), and V(V), and oxalic acid to mask Ti(IV).[58] Quinaldinic acid (with masking by tartaric acid and ammonium chloride) has also been proposed as a precipitant for Pd(II).[43] Addition of hydrogen peroxide to dilute hydrochloric acid solutions containing tartaric acid converts Ti, Nb, and Ta to peroxy complexes which do not interfere when Zr is precipitated with α-hydroxybenzylphosphonic acid.[59] If fluoride is present it can be masked by boiling with boric acid.

A mixture of tartrate (to mask As(V)), citrate (for Ag, Pb(II), Bi(III), Sb(III), and Th(IV)), oxalate (for Sn(IV)), and fluoride ion (for Ti(IV) and Zr(IV)), was used when Hg(II) was determined gravimetrically with *N*-benzoyl-*N*-phenylhydroxylamine at pH 3–6.[60] Chloride, cyanide, and EDTA caused interference.

When a silicon–germanium alloy is dissolved in nitric/hydrofluoric acid mixture and the silicon is removed by volatilization as SiF_4, much of the Ge may also be lost. If the process is carried out in the presence of oxalic acid, the Ge is masked against the formation of GeF_4 and is retained, so that it can be ignited to GeO_2 and determined gravimetrically.[61]

Masking of Hg(II) with cyanide ion in ammoniacal solution permits the precipitation of Tl(I) as Tl_2CrO_4 in the presence of Hg(II).[62] Cyanide ion also masks Ag, Cu, Cd, Hg, Co, Ni, Fe, and Zn in the gravimetric determination of Bi with 2-amino-1,3,4-thiadiazole-5-thiol.[63] Fluoride ion masks Pb, Bi, and Fe(III).[63] At pH 6.4–7.2, La can be precipitated with *N*-benzoyl-*N*-phenylhydroxylamine, in the presence of cyanide ion to mask Ni, Cu, Zn, and Fe(II) (by reduction of Fe(III) with sulfite ion).[64]

8-Hydroxyquinoline cannot be used as a masking agent because it is too sparingly soluble. This disadvantage can be overcome by inserting a 5-sulfonic acid group to confer water solubility without affecting the chelating ability of the molecule. This makes it possible to use 8-hydroxyquinoline-5-sulfonic acid to retain other metal ions in solution when calcium is precipitated from ammonia buffer at pH 9 as its oxalate.[65]

HOMOGENEOUS PRECIPITATION

In the technique of precipitation from homogeneous solution[66,67,68,68a] the concentration of one of the components of the species to be precipitated is kept low, but is slowly increased. The resulting precipitate is often an extremely fine one and is less likely to adsorb other species. The process may be one of anion release, as in the liberation of sulfate ions by hydrolysis of dimethyl sulfate,[69] sulfamic acid[70] or ammonium persulfate,[71] and of sulfide ion by hydrolysis of thioacetamide,[71a] thiourea, or thioformamide. Similarly, cation release may be involved, by destruction of a complexing species, by oxidation, hydrolysis or volatilization (ammonia). Again, the process may be one of slow pH change, such as by the hydrolysis of urea with the liberation of ammonia,[71b] or of change in the valence state of the metal ion.[71c,71d,71e] Cation replacement is sometimes possible, as in the precipitation of barium sulfate from a barium–EDTA solution by adding a nickel solution and sulfate ions.[71f]

Masking can sometimes be used to decrease the concentrations of other metal ions to sufficiently low levels to prevent interference while still permitting the precipitation of the desired species; for example, EDTA has been added as a general masking agent in the homogeneous precipitation of Ba as $BaSO_4$. Ammonium persulfate was added to the solution at pH 10–11, and on warming, the persulfate slowly hydrolyzed to sulfate and hydrogen ions. The pH slowly fell, progressively demasking Ba from BaEDTA while other metal ions remained complexed, and with increasing sulfate concentration, $BaSO_4$ was quantitatively precipitated.[71] Similarly, EDTA improves the selectivity that can be achieved in the homogeneous precipitation of metal ions as their sulfides, based on hydrolysis of thioacetamide.[13,72] In the homogeneous precipitation of Ti(IV) from slightly acid solution with cupferron, tartaric acid is used to mask Al, borate, phosphate, and vanadate.[73] In alkaline solution ammonia masks silver against precipitation as silver chloride, but the pH of the solution can be lowered by the slow hydrolysis of β-hydroxyethylacetate and silver chloride is precipitated.[72a]

Beryllium is precipitated from homogeneous solution by decomposition of its acetylacetonate (formed initially in acid solution) when the solution is made alkaline with ethylenediamine and heated.[74] In this method, Cu(II) is masked by the ethylenediamine, and Fe, Ca, and Mg are masked by the addition of a slight excess of EDTA.[74]

An interesting type of precipitation from homogeneous solution is based on a photochemical reaction. Tantalum is masked by oxalate ion against precipitation as selenite, but the oxalate is destroyed by indirect catalytic oxidation by photochemically produced bromine.[75] The reaction is not quantitative. Photochemical reduction of periodate has been used to generate iodate as a precipitant for Th(IV).[76]

EDTA has been used as a masking agent to improve the fractionation of pairs of rare earths by homogeneous precipitation as their oxalates, based on hydrolysis of methyl oxalate.[77]

Ferric ion has been precipitated at pH 3.0 as its hydroxide, from EDTA solution, by controlled destruction of the ligand with hydrogen peroxide at 30°.[78] Ferric and ferrous ions can also be precipitated at pH 2–6 as their tris(2-thiopyridine-*N*-oxide) complexes, the ligand being formed by slow hydrolysis of *S*-2-pyridylthiouronium bromide 1-oxide in the hot dilute solution. Selectivity is improved by including such masking agents as citrate, thiourea and hydroxylamine (as a reducing agent).[79]

The complexing species can sometimes be synthesized in solution. In the precipitation of Ni salicylaldoximate from homogeneous solution at pH 5.9–6.8, the ligand is synthesized in solution from hydroxylamine and salicylaldehyde. Addition of citric acid and EDTA prevents the hydrolytic precipitation of Ni, and tartaric acid masks Fe and Zn.[80]

If Cd is present, tartrate cannot be used because Cd tartrate is only sparingly soluble. Other examples involving the synthesis of ligands directly in the solution have been reviewed.[68a]

The advantages of homogeneous precipitation at constant temperature and pH lie in the slow rate of crystal growth and in the much closer approximation to equilibrium between the solid phase and the solution, thereby minimizing adsorption, occlusion, coprecipitation, and other undesirable effects. On the other hand, this method can be used for only a small number of anions (which do, however, include oxalate, phosphate, and sulfate, formed by hydrolysis of their alkyl esters), and it cannot always be assumed that the resulting precipitates will be uniform and crystalline. When closely similar cations are present in solution, coprecipitation may occur. Also, in situations in which more than one kind of ligand is present, for example, in ligand displacement reactions, mixed-ligand complex formation must be considered. Thus, when a **1:2** ratio of rare earths and EDTA was treated at pH 4.5–4, at room temperature, with an equivalent amount of various ligands as precipitants, species of the types MYA were obtained, where M is a rare earth metal, Y is the anion of EDTA, and A is oxalate, carbonate, phosphate or fluoride ion.[81]

REFERENCES

1. K. Pamnani, *Z. Anal. Chem.* **222,** 399 (1966).
2. G. W. C. Milner, and J. W. Edwards, *Anal. Chim. Acta* **13,** 230 (1955).
3. A. Ringbom, *Complexation in Analytical Chemistry,* Wiley, New York, 1963, Chap. 3.
4. K. Wang, *Acta Chim. Sinica* **29,** 395 (1963).
5. W. P. Han, *Acta Chim. Sinica* **30,** 422 (1964).
6. K. Wang, *Acta Chim. Sinica* **30,** 250 (1964).
7. Ref. 4, p. 78.
8. C. S. Wang, W. P. Han, and K. Wang, *Acta Chim. Sinica* **31,** 476 (1965).
9. R. Pribil, and V. Vesely, *Talanta* **10,** 233 (1963).
10. R. Pribil, and V. Vydra, *Coll. Czech. Chem. Comm.* **24,** 3103 (1959).
11. I. L. Teodorovich, and R. Gutnikova, *Zh. Analit. Khim.* **21,** 1127 (1966).
12. K. N. Johri, and K. Singh, *Curr. Sci. India* **34,** 582 (1965).
12a. K. N. Johri, N. K. Kaushik, and K. Singh, *Talanta* **16,** 432 (1969).
13. K. L. Cheng, *Anal. Chem.* **33,** 783 (1961).
14. R. Pribil, and M. Kopanica, *Chemist-Analyst* **48,** 87 (1959).
15. G. Lanfranco, and F. Cerrato, *Anal. Chim. Acta* **39,** 47 (1967).
16. G. Lanfranco, *Anal. Chim. Acta* **38,** 523 (1967).
17. A. Y. Sheskol'skaya, *Zh. Analit. Khim.* **21,** 1138 (1966).
18. R. Pribil, *Coll. Czech. Chem. Comm.* **16,** 542 (1951).
19. R. Pribil, and V. Sedlai, *Coll. Czech. Chem. Comm.* **16,** 69 (1951).
20. See, for example, R. Pribil, *Chem. Listy* **45,** 57 (1951).

21. A. Hulanicki, *Talanta* **9,** 549 (1962).
22. J. J. Kelly, and D. C. Sutton, *Talanta* **13,** 1573 (1966).
22a. D. E. Ryan, *Talanta* **15,** 486 (1968).
23. G. Gottschalk, and P. Dehmel, *Z. Anal. Chem.* **212,** 380 (1965).
24. A. K. Sengupta, *J. Indian Chem. Soc.* **41,** 707 (1964).
25. S. Misumi, and T. Taketatsu, *Bull. Chem. Soc. Japan* **32,** 593 (1959).
26. R. G. Monk, and K. A. Exelby, *Talanta* **12,** 91 (1965).
27. A. K. Sarkar, and J. Das, *Anal. Chem.* **39,** 1608 (1967).
28. R. Pribil, and K. Kucharsky, *Coll. Czech. Chem. Comm.* **15,** 132 (1950).
29. R. Pribil, and P. Schneider, *Coll. Czech. Chem. Comm.* **15,** 886 (1950).
30. J. Hure, M. Kremer, and F. le Berquier, *Anal. Chim. Acta* **7,** 37 (1952).
31. V. G. Goryushina, *Zavodskaya Lab.* **21,** 148 (1955).
32. R. Pribil, and J. Adam, *Coll. Czech. Chem. Comm.* **18,** 305 (1953).
33. P. I. Brewer, *Analyst* **77,** 539 (1952).
34. R. Pribil, and L. Fiala, *Chem. Listy* **46,** 331 (1952).
35. I. A. Tarkovskaya, and F. P. Gorbenko, *Zh. Analit. Khim.* **20,** 1185 (1965).
36. V. I. Spitsyn, K. B. Zaborenko, and S. A. Brusilovskii, *Zh. Neorg. Khim.* **1,** 2160 (1956).
37. A. Claassen, L. Bastings, and J. Visser, *Analyst* **92,** 618 (1967).
38. K. L. Cheng, *Chemist-Analyst* **50,** 126 (1961).
39. L. Erdey, I. Buzas, and K. Vigh, *Talanta* **14,** 515 (1967).
40. R. Ripan, and V. Sacelean, *Talanta* **12,** 69 (1965).
41. R. Pribil, and J. Vorlicek, *Chem. Listy* **46,** 216 (1952); *Coll. Czech. Chem. Comm.* **18,** 304 (1953).
42. A. K. Majumdar, and S. P. Bag, *Z. Anal. Chem.* **165,** 247 (1959).
43. A. K. Majumdar, and J. G. Sen Gupta, *Z. Anal. Chem.* **161,** 100, 104 (1958).
44. A. K. Majumdar, and S. P. Bag, *Z. Anal. Chem.* **164,** 394 (1958).
45. S. Prakash, R. P. Singh, and K. C. Trikha, *Talanta* **13,** 1393 (1966).
46. W. J. Holland, J. Bozic, and J. Gerard, *Analyst* **93,** 490 (1968).
47. A. I. Busev, and M. I. Ivanyutin, *Zh. Analit. Khim.* **13,** 18 (1958).
48. B. C. Bera, and M. M. Chakrabartty, *Z. Anal. Chem.* **223,** 169 (1966).
49. M. Malinek, *Chem. Listy* **49,** 1400 (1955).
50. V. Sedivec, *Chem. Listy* **45,** 177 (1951); *Coll. Czech. Chem. Comm.* **16,** 398 (1951).
51. P. F. Lott, and R. K. Vitek, *Anal. Chem.* **32,** 392 (1960).
52. J. Bassett, G. B. Leton, and A. I. Vogel, *Analyst* **92,** 279 (1967).
53. B. Das, and S. C. Shome, *Anal. Chim. Acta* **40,** 338 (1968).
54. C. A. Starace, L. D. Wiersma, and P. F. Lott, *Chemist-Analyst* **55,** 74 (1966).
55. P. F. Lott, P. Cukor, G. Moriber, and J. Solga, *Anal. Chem.* **35,** 1159 (1963).
56. S. Hikime, and L. Gordon, *Talanta* **11,** 851 (1964).
57. B. R. Singh, and S. Kumar, *Indian J. Chem.* **3,** 410 (1965).
58. S. N. Poddar, *Anal. Chim. Acta* **28,** 586 (1963).
59. J.-Y. Yen, C.-H. Yung, and H.-Y. Liu, *Hua Hsueh Hsueh Pao* **30,** 562 (1964).
60. B. Das, and S. C. Shome, *Anal. Chim. Acta* **35,** 345 (1966).

61. K. L. Cheng, and B. L. Goydish, *Anal. Chem.* **35,** 1273 (1963).
62. I. M. Korenman, *Analytical Chemistry of Thallium,* Israel Programme for Scientific Translations, Jerusalem, 1963, p. 51.
63. E. Domagalina, and L. Przyborowski, *Chem. Anal.* (*Warsaw*) **11,** 1087 (1966).
64. B. Das, and S. C. Shome, *Anal. Chim. Acta* **32,** 52 (1965).
65. A. D. Maynes, *Anal. Chim. Acta* **32,** 288 (1965).
66. H. B. Willard, *Anal. Chem.* **22,** 1372 (1950).
67. F. H. Firsching, *Talanta* **10,** 1169 (1963).
68. L. Gordon, M. L. Salutsky, and H. B. Willard, *Precipitation from Homogeneous Solution,* Wiley, New York, 1959.
68a. P. F. S. Cartwright, E. J. Newman, and D. W. Wilson, *Analyst* **92,** 663 (1967).
69. P. J. Elving, and R. E. Van Atta, *Anal. Chem.* **22,** 1375 (1950).
70. W. F. Wagner, and J. A. Wuellner, *Anal. Chem.* **24,** 1031 (1952).
71. A. H. A. Heyn, and E. Schupak, *Anal. Chem.* **26,** 1243 (1954).
71a. W. C. Broad, and A. J. Barnard, *Thioacetamide as a Sulfide Precipitant,* J. T. Baker Chemical Co., New Jersey, 1960.
71b. H. H. Willard, and N. K. Tang, *J. Amer. Chem. Soc.* **59,** 1190 (1937); *Ind. Eng. Chem., Anal. Ed.* **9,** 357 (1937).
71c. H. H. Willard, and S. T. Yu, *Anal. Chem.* **25,** 1754 (1953).
71d. W. A. Hoffman, and W. W. Brandt, *Anal. Chem.* **28,** 1487 (1956).
71e. E. J. Newman, *Analyst* **88,** 500 (1963).
71f. F. H. Firsching, Dissertation, Syracuse University, 1954.
72. H. Flaschka, *Chemist-Analyst* **44,** 2 (1955).
72a. L. Gordon, J. I. Peterson, and B. P. Burtt, *Anal. Chem.* **27,** 1770 (1955).
73. A. H. A. Heyn, and N. G. Dave, *Talanta* **13,** 33 (1966).
74. W. G. Boyle, and C. H. Otto, *Anal. Chem.* **39,** 1647 (1967).
75. J.-Y. Yen, and W. Yang, *Sci. Sinica* (*Peking*) **13,** 343 (1964).
76. M. Das, A. H. A. Heyn, and M. Z. Hoffman, *Talanta* **14,** 439 (1967).
77. L. Gordon, and K. J. Shaver, *Anal. Chem.* **25,** 784 (1953); and references therein.
78. W. M. MacNevin, and M. L. Dunton, *Anal. Chem.* **26,** 1246 (1954).
79. J. A. W. Dalziel, and M. Thompson, *Analyst* **89,** 707 (1964).
80. B. S. K. Rao, and O. E. Hileman, *Talanta* **14,** 299 (1967).
81. S. Pajakoff, *Ost. Chem.-Zeit.* **67**(2), 42 (1966).

CHAPTER

7

SPECTROPHOTOMETRY

In spectrophotometric methods based on complex formation the commonest cause of interference comes from the reaction of one or more foreign ions, either with the species being determined or, more frequently, with the organic reagent. If a suitable masking agent can be found, this interference can be overcome without having to introduce a preliminary separation step into the method. In general, masking requirements are less rigid than in complexometric titrations, especially if the colored species is finally extracted into an organic solvent, because spectrophotometric methods are usually, by their nature, more selective. The order of addition of reagents is sometimes important, because the rate of formation or dissociation of an unwanted complex, either of the color-forming species with an interfering metal ion or of the masking agent with the metal to be determined, may be very slow. The latter is more likely to be the case if the masking agent is a multidentate ligand such as EDTA. Thus the kinetically inert complex of Al with 5-sulfo-4′-diethylamino-2′,2-dihydroxyazobenzene is attacked only slowly by EDTA, so the latter, added after color development, can be used as a masking agent for Cr, Cu, Th, Ti, U, V, and W in the spectrophotometric determination of Al with this reagent.[1] For a masking agent to be satisfactory for use in spectrophotometry, it must meet a number of requirements.

In cases where determinations are carried out directly on the solution it should not, itself, be strongly colored and neither should the complexes that it forms. Sulfosalicylic acid is a suitable masking agent for Al in the photometric determination of Ce and the rare earths with arsenazo III in acetate buffer, but if Fe(III) is present a strongly colored complex is formed. In this case the difficulty can be avoided, and at the same time Cu can be masked, if thiourea is also added to reduce Fe(III) to Fe(II).[2]

If the species being determined is finally extracted into an organic solvent, colored complex formation by a masking agent is not important so long as these complexes are not also extracted.

A masking agent should not form very stable complexes with the species that is to be determined. However, in some cases, gradation in masking ability can be secured by selective displacement. Calcium, but not strontium, interferes in the spectrofluorimetric determination of Mg with *N*,*N'*-bis-salicylidene-2,3-diaminobenzofuran at pH 10.5. EDTA cannot be used to mask Ca, because it would also mask Mg, but Sr–EDTA is satisfactory for the same purpose because the stability constants of the EDTA complexes lie in the sequence Ca > Sr > Mg.[3] The Sr–EDTA complex also masks La and Y.

It should not react chemically with the reagent used to develop the colored species, nor should it oxidize or reduce the metal ion that it is desired to estimate. Hydrogen peroxide would not be satisfactory as a masking agent in the determination of Fe(II) with 1,10-phenanthroline, nor would thiourea in the determination of Fe(III) with thiocyanate ion.

It should, however, form stable complexes with the interfering ions or else oxidize or reduce them to valence states in which they no longer cause trouble.

Similarly, all masking agents that are used must be compatible with one another and, by making a suitable selection, it should be possible to eliminate all serious interferences. This is not always easy, especially if the foreign metal ion and the species to be determined have comparable affinities for the reagent used. Copper interferes seriously in the determination of Te with bismuthiol II, and is not masked by EDTA, ascorbic acid, or oxalate ion. This difficulty is to be expected because of the structural similarity of bismuthiol II to the diethyldithiocarbamate ion which reacts with Cu even in the presence of EDTA.[4]

Typical examples of the applications of masking agents in spectrophotometry, either singly or in admixture, are described below.

MASKING WITH EDTA

The ability of EDTA to form complexes with most metal ions, and the dependence of this property on the pH of the solution, together make EDTA one of the most useful of the more general masking agents. It also has the advantage that almost all of its metal complexes are readily soluble and either colorless or only weakly colored. Its main applications as a masking agent are in determinations of metal ions that complex with it only relatively weakly. Examples include Ag, Be, B, V, Nb, Pd, Sb, and Se. EDTA can also be used satisfactorily as a masking agent in the determination of other metal ions, so long as their EDTA complexes are relatively less stable than the complexes formed with the reagents used for the determinations. For example, Cu and Bi can be determined with sodium diethyldithiocarbamate in the presence of EDTA as masking agent. Similarly, in the spectrophotometric determination of Ni with 4-(2-pyridylazo)resorcinol at pH 8.6–10, EDTA masks Al, Cd, Cr, Cu, Ga, Hg, In, Pb, V, Zr, and the rare earths.[5]

When Be is determined with 2-phenoxyquinizarin-3,4′-disulfonic acid at pH 6.0, moderate amounts of many cations, including Al, are masked by CaEDTA.[6] Similarly, for the determination of Be in Be–Cu alloys, using aluminon in ammonium acetate buffer, EDTA was added to mask Al, Co, Cu, Fe, Mn, Ni, Ti, Zn, and Zr.[7] Use of EDTA allowed Be in Al to be determined with sulfosalicylic acid (at pH 10).[8] Nitrilotriacetic acid and EDTA were good masking agents in the spectrophotometric determination of Be with xylenol orange in pH 5.8 buffer[9] or with beryllon IV.[10] At pH 7–8, EDTA masked Bi, Cd, and Cr.

Conversely, at lower pH values Al was less effectively masked so that, at pH 3–3.8, xylenol orange could be used to determine Al, whereas EDTA continued to mask Fe (reduced by ascorbic acid), Ce, Co, La, Mg, Mo, Nd, and Pb.[11,12]

In the photometric determination of B in aqueous solution with pyrocatecholphthalein, at pH 9–9.5, bivalent cations can be masked by EDTA, but ter- and tetravalent ones must be removed (for example by ion exchange).[13]

Thorium can be determined by extraction from 0.6*M* hydrochloric acid as a metal chelate with the dianilide of 3,6-bis(2′-arsonophenylazo)-4,5-dihydroxy-2,7-naphthalenedisulfonic acid, using cyclohexanediaminetetraacetic acid (CDTA) to mask actinides, lanthanides, Zr, Hf, Y, and Sc.[14] At pH 10, EDTA masks Ca, Sr, Ba, Mn, Co, Cu, Hg, Al, Sn, Pb, Fe, U, and the rare earths when Th is determined with thymolphthalexon.[15] In both of these examples, masking probably depends on the stereochemistry of the metal ions concerned. With common, six-coordinate metal ions EDTA and CDTA are well suited to form octahedral complexes that are more stable than the complexes of thymolphthalexon. However, Th is eight-coordinate in its complexes, such as the tetrakis acetylacetonate, which has a square antiprism structure, and although it forms a stable complex with EDTA, its steric requirements appear to be better met by tetradentate complex formation involving two oxygens of carboxy anions, the corresponding nitrogen, and the phenolic oxygen of thymolphthalexon.

Vanadium has been determined spectrophotometrically in uranium materials by extraction of its 8-hydroxyquinolinate into carbon tetrachloride. At pH 3.5–4.5, EDTA masks many cations, especially Fe, and excess Th has to be added to displace any V that is also complexed by the EDTA.[16] These conditions are near the limit for satisfactory masking of V and, in fact, EDTA cannot be used as a masking agent in the colorimetric determination of V(V) with 4-(2-pyridylazo)resorcinol (PAR) at pH 5.5 because it breaks down the V-PAR complex.[17] On the other hand, CDTA is satisfactory, forming complexes under these conditions with most metal ions but not with V, so that the method becomes almost specific.[17] A similar improvement in selectivity was found in the colorimetric determination of V with xylenol orange.[18]

4-(2-Thiazolylazo)resorcinol reacts with Nb in acetate–tartrate medium (pH 5–6) to form a colored complex. From the chemical similarities with the reaction between V and PAR, interference from V is to be expected unless it is reduced to VO^{2+} by hydroxylamine or saturated ammonium oxalate, but the use of EDTA or CDTA masks most other cations, such as Fe(III), Al, Zn, Mg, and Mn. (Mo, Ti, U, and Th are not masked.)[19]

However, the equilibria may be more complicated. At pH 5.8 in a tartrate medium, Nb and PAR react to form a ternary complex of Nb, PAR, and tartrate. With EDTA as masking agent, only U, V, and phosphate ion interfere.[20] Bromopyrogallol can be used instead of PAR.[20]

In the photometric determination of Nb with sulfochlorophenol C in *M* hydrochloric acid, EDTA or potassium fluoride masks Zr and Hf.[21] A selective test for Nb is based on the formation of an intense blue color with methylthymol blue at pH 5 in the presence of hydrogen peroxide: many metals including Mo, Ta, Th, Ti, W, and Zr are masked by CDTA, but Cr forms a blue Cr–H_2O_2–CDTA complex.[22]

Manganese(III) and Fe(III) form colored complexes with formaldoxime, but by adding hydroxylamine and EDTA the iron complex is dissociated and the iron is reduced and masked as Fe(II)–EDTA, so that Mn can be determined colorimetrically.[23] Selectivity can be improved in the spectrophotometric determination of Fe as Fe(II) if cations are masked with EDTA in acid solution before the pH is adjusted to 7–8, and Fe(III) is reduced to Fe(II) with hydroxylamine in the presence of 1,10-phenanthroline:[24] use of EDTA in this way also prevents the hydrolysis of Fe(III). A similar method for Cu adds sodium diethyldithiocarbamate to the EDTA solution at pH 9, followed by extraction into chloroform.[24]

Whereas Co is stabilized as Co(III) in its complex with 6-amino-4-hydroxy-2-mercapto-5-nitrosopyrimidine, Ni and Cu are masked by EDTA at pH 9.2: this is the basis of a spectrophotometric method for Co.[25] Nickel and Cu are also masked by EDTA in a method for the determination of Pd in alkaline aqueous ethanol, using pyridine-2-aldehyde-2-pyridylhydrazone (PAPHY).[26]

EDTA has been used as a "mass-masking agent" for most metal ions in the spectrophotometric determination of Cu(II) with acid alizarine black SN in aqueous ammonia at pH 11.2.[27] It is not satisfactory for masking Th, Bi, La, Fe(III), or Co(II), and indeed, acid alizarine black SN has also been proposed as a reagent for Th.[28] The weakness of the Ag complex with EDTA makes EDTA a good masking agent for use in the spectrophotometric determination of Ag by mixed complex formation with bromopyrogallol red and 1,10-phenanthroline.[29] Thus EDTA masks metal ions such as Pb(II) which would react with bromopyrogallol red, Fe(II) (by oxidation to Fe(III)) which would react with 1,10-phenanthroline, or Hg(II) which would react with both. (If U(VI) or Th(IV) were present, they could be masked by fluoride ion, and Nb(V) could be masked by hydrogen peroxide.)[29]

The Hg–dithizone complex is more stable than Hg–EDTA, so that EDTA can be used to mask most metals in the colorimetric determination of Hg by dithizone extraction. One of the few interferences is Ag, which is masked by thiocyanate ion.[30] The same comments apply to the use of 1,5-di(β-naphthyl)thiocarbazone, instead of dithizone, to determine Hg as an impurity in Se.[31]

Almost all interfering cations are masked by EDTA in the photometric determination of Sb(V) in dilute hydrochloric acid with pyrocatechol violet,[32] and of Se(IV) as diphenylpiazselenol, formed with 3,3′-diaminobenzidine and extracted into toluene.[33] Similarly, in the spectrophotometric determination of Se in technical sulfuric acid by piazselenol formation with *o*-phenylenediamine, Fe(III) is masked by EDTA and nitrous acid is masked by urea.[34]

Finally, EDTA, itself, can be determined colorimetrically by its masking effect on a known excess of Fe(III) in the presence of sulfosalicylic acid or thiocyanate.[35] Other alternatives are to use excess Ni with dimethylglyoxime,[36] or Zn with zincon[37] or 8-hydroxyquinoline. EDTA has been determined spectrophotometrically in urine by reduction of chromate to Cr(III) with the formation of Cr(III)–EDTA: addition of Zn(II) to a duplicate sample masks the reaction and serves as a control.[38]

MASKING WITH CARBOXYLIC ACIDS

In most cases, carboxylic acids and their hydroxy derivatives bind metal ions rather weakly, but the stability of the complexes is appreciable if the cations are polyvalent. Thus, Zr can be masked by oxalic acid when Th is determined with arsenazo III in 2.5–3M hydrochloric acid.[39] In the absence of Th, U(IV) can be determined in the same way in 3–9M hydrochloric acid.[39] If Fe(III) is present it must be reduced with ascorbic acid. Similarly, tungsten is masked by tartaric acid in the photometric determination of Mo in steels by mercaptoacetate extraction from *M* hydrochloric acid into isoamyl alcohol.[40] Oxalate ion is used to mask molybdenum in the diphenylcarbazide test for chromium.[40a]

The tris 1,10-phenanthroline–Fe(II) complex is much more stable than the Fe(II)–oxalate complexes, so that Fe can be determined in this way in oxalic acid and metal oxalate solutions, if ascorbic acid is added to reduce Fe(III).[41] Citrate is used as a buffer (pH 7) and also to mask Cr, Th and Zn in a method for the spectrophotometric determination of Fe with ethylenediamine-di(*o*-hydroxyphenylacetic acid).[42]

The simultaneous determination of Ag and Cu in high-purity lead by spectrophotometric extractive titration at pH 4.3–5.5 with dithizone (after precipitating most of the lead with ammonium sulfate) avoids interference from the following ions by masking them with tartrate ion: Bi, Cd, Co, In, Mn, Ni, Pb, and Tl. Zinc is not adequately masked in this way.[43] With ammonium tartrate in a borate buffer (pH 6), as masking agent, the following spectrophotometric determinations of metals in tungsten alloys were possible: Fe with 1,10-phenanthroline in the presence of hydroxylamine, Ni with dimethylglyoxime, Cu with neocuproine in the presence of hydroxylamine, and Co with nitroso-R-salt.[44] The copper content of impure gallium was determined using neocuproine, with citrate as buffer (pH 4–6) and masking agent for Ga.[45]

Citrate has been used to mask Te in the spectrophotometric determination of Cu in Te, with 2,9-dimethyl-1,10-phenanthroline,[46] and in the determination of Fe(II) with 1,10-phenanthroline in Th and its compounds and in uranium alloys containing Ti, Mo, and Al.[47] At pH 3 it also masks Sb, Sn, Te, and Tl.[48] Iron in Ta metal has been determined using tartaric acid as masking agent.[47]

A mixture of potassium tartrate and potassium citrate in ammonium chloride/ammonia buffer was used as a masking agent when Cd was extracted as its diethyldithiocarbamate into chloroform.[49]

In the spectrophotometric determination of tin in steel, by extraction of its phenylfluorone complex into methyl isobutyl ketone from dilute hydrochloric acid, oxalic acid masked Fe(III), Hf(IV), Nb(V), Ta(V), V(V), W(VI), and Zr(IV). By including hydrogen peroxide, Mo(VI), and Ti(IV) were also masked.[50] Tartaric acid masked Sb, W, and Zr when tin was determined with gallein.[51] If Fe(III) or V(V) were present, they could be reduced by ascorbic acid. Molybdenum and Ti were masked by adding hydrogen peroxide.[51]

The difficulty in laying down firm guiding principles for qualitative choice of masking agents is seen in the following example. When pyrogallol-4-sulfonic acid in strongly acid medium was used for the spectrophotometric determination of Nb and Ta, citric acid masked Ta(V), but not Nb(V), whereas oxalic acid masked Nb(V) but not Ta(V).[52]

MASKING WITH CYANIDE ION

The preference of cyanide ion is for complex formation with the more polarizable cations, usually of low charge. Thus, in the determination of La(III) with xylenol orange in aqueous pH 7.5 buffer, cyanide ion masks Ag, Cd, Co(II), Cu, Ni, and Zn.[53] The iron content of Cu-rich materials has been determined by color formation with 4,7-diphenyl-1,10-phenanthroline (bathophenanthroline) using cyanide ion to mask Cu as $Cu(CN)_4^{3-}$, and also Zn as $Zn(CN)_4^{2-}$.[54] Conversely, in the determination of Al in steel, using 8-hydroxyquinoline, Fe(III) can be masked by reduction and adding cyanide ion to form ferrocyanide.[55]

Masking of Cu with cyanide ion was also used when Co(II) was determined in ammoniacal buffer with 3-mercaptopropionic acid.[56] Similarly, in the determination of Pb(II) with 4-(2-pyridylazo)resorcinol in a pH 10 ammonium chloride/ammonia buffer, cyanide ion masked Ag, Cd, Co, Cu, Hg, Ni, and Zn.[57] The avidity of cyanide ion for Hg(II) has been proposed as a method for determining cyanide ion, based on its masking action towards Hg(II) present in solution as its *p*-dimethylaminobenzylidenerhodanine complex.[58]

MASKING WITH HALIDE IONS

Fluoride ion is useful as a masking agent mainly for polyvalent cations such as Al(III), Fe(III), and Th(IV). At pH 2 fluoride ion masks Th(IV) but not UO_2^{2+}, which can be determined with arsenazo III.[59] In the determination of V(IV) with xylenol orange at pH 2.8, fluoride masks Al, Cr, Sn, Th, Ti, Y, and Zr.[60] Similarly, Fe(III) and Ti are masked in the spectrophotometric determination of V(V) with EGTA and hydrogen peroxide.[61] Iron(III) can be masked by adding ammonium fluoride when Co(II) is determined using 2-nitroso-1-naphthol,[62] and fluoride has also been suggested for masking Fe(III) in an Fe(II), Fe(III) mixture when Fe(II) is determined with 1,10-phenanthroline.[63] For the determination of Mo in uranium, using xylenol orange in an acetate buffer, ammonium fluoride masks Al, Bi, Fe, In, Sc, Th, Ti, Y, and Zr.[64] At higher concentrations, the range of its masking activity is extended, so that in a spot test for Cu using solochrome azurine BS, Co(II), Ni, U(VI), and V(IV) are also masked.[65]

Conversely, fluoride interferes in many spectrophotometric methods, for example the determination of Nb with thiocyanate ion. In this case it can be masked by adding Al as its tartrate to form AlF^{2+}, AlF_2^+, and AlF_3.[66] Spectrophotometric methods for fluoride ion often depend on its masking ability towards polyvalent cations thereby displacing, for example, metals such as Th from Th–alizarine red S complex[67] or Fe(III) from Fe(III)–acetylacetone complex,[68] with resulting decrease in color.

Iodide ion masks Cd(II) when Ga(III) is determined with PAN at pH 5–6.[69] Small amounts of Hg in water can be determined using dithizone, with potassium thiocyanate to mask Ag.[70]

MASKING WITH SULFUR-CONTAINING LIGANDS

Bis(carboxymethyl)dithiocarbamate has been proposed as a selective masking agent for Cu, Pb, Cd, and Ni in the extraction of Zn dithizonate.[71] The stability sequence for the closely related bis(2-hydroxyethyl)dithiocarbamate (made by dropwise addition of carbon bisulfide to diethanolamine in methanol[72]) is: Au(III) $>$ Hg $>$ Ag $>$ Cu $>$ Bi $>$ Pb $>$ Fe(III) $>$ Cd $>$ Co $>$ Ni $>$ Zn $>$ Mn(II). In this series, all metals to the left of Zn, and also Sn(II) and Tl(I), are masked by this reagent when Zn in $GeCl_4$ and GeO_2 is determined by dithizone extraction from a tartrate solution of pH 8–9.[72] The tartrate masks Ge against hydrolytic precipitation, and progressive increments of dithizone are used to avoid large excesses which might begin to extract Cu and Cd. This dithiocarbamate has also been supplemented with cupferron to remove Fe and Cu in the dithizone method for Zn.[73] It has been used as a reagent for the photometric determination of Cu.[74] Addition of Pb did not affect the reaction but masked the formation of colored complexes by Co, Mn, and Ni. Interference by Fe(III) was avoided by adding fluoride ion or by reduction to Fe(II) with ascorbic acid, hydroxylamine hydrochloride, or hydrazine sulfate.[74]

Thioglycolic acid masks Co and Fe in the spectrophotometric determination of Al with aluminon at pH 5.3.[75] It has been used as a masking agent for Cu at pH 9[47] and also to mask Cd, Co, Cu, Hg, and Zn in a method where Ni in sea water is determined by extraction of its pyridine-2-aldehyde-2-quinolylhydrazone complex into benzene.[76]

OTHER MASKING AGENTS

Masking with phosphate is sometimes useful for polyvalent cations such as Fe and Ce. This technique was applied in the spectrophotometric determination of Mn by extraction with 2-thenoyltrifluoroacetone,[77] and to mask Fe(III) when V was determined as its 5,7-diiodo-8-hydroxyquinoline complex.[78] Pyrophosphate is also suitable.[79] A spot test for Hg(II) and Ag, using dithizone, is based on the presence of pyrophosphate as a masking agent for most cations.[80]

Sometimes, however, none of these masking agents is satisfactory. In the determination of Nb with pyrocatechol violet at pH 1.3–2.2, EDTA or phosphate is unsuitable because of reaction with Nb, nor can citric, tartaric, or oxalic acids be used. In this case mannitol masks Ti, Al, V, W, and Mo.[81]

Xylenol orange is not very selective as a colorimetric reagent and it has, indeed, been proposed for the photometric determination of many metals, including Al, Be, Hf, Mo, Pb, V, Zn, and Zr. At pH 5.6, microgram quantities of Zn and Pb can be determined with this reagent, in the presence of each other, if, in one case, Pb is masked with sodium thiosulfate and, in the other case, Zn is masked with potassium ferrocyanide.[82] Triethanolamine or lactic acid is a suitable masking agent for Al when Mg is determined photometrically with titan yellow, phenazo, or magneson-2.[82a] Hexamethylenediaminetetraacetic acid is another possibility.

In the presence of sulfate ion, hydrogen peroxide masks Zr, but not Hf, against complex formation with xylenol orange in 0.1–0.2*M* perchloric acid, so that Zr and Hf can be determined individually and in admixture.[83] Tiron (4,5-dihydroxybenzene-1,3-disulfonic acid) has been used at pH 10 to mask Fe and Cu when Ni was extracted as its PAN complex.[84]

MIXTURES OF MASKING AGENTS

More usually, however, two or more masking agents of quite different types are used at the same time to limit even further the range of possible interferences. One such combination is EDTA with citric or tartaric acid. A citrate–EDTA mixture at pH 5–6.5 was used in a selective spectrophotometric determination of Fe(II) with 1,10-phenanthroline. It masked Al, Bi, Cd, Cr, Mo, Sb, Sn, Th, Ti, U, W, Zn, and Zr. By including mercaptoacetic acid and tartaric acid, Cu and Ta were also masked.[85] Ammonium citrate and EDTA masking are used in the extraction of copper (II) as its diethyldithiocarbamate. Similarly in a colorimetric method for Ag, based on an exchange reaction between Ag^+ and the Cu complex of tetraethylthiuram disulfide (disulfiram), selectivity was increased by adding EDTA and tartaric or citric acids to mask Bi, Pb, Sb, and Sn.[86] In alkaline solution, triethanolamine and tartrate masked Fe, Ni, and V in the photometric determination of U with hydrogen peroxide.[87] Triethanolamine and EDTA were used as masking agents when Be was determined with beryllon(III).[88] Citric and thioglycolic acids were used to mask Al, Cd, Cu, Fe(III), Hg, and Zn when Ni was extracted into chloroform as its pyridine-2-aldehyde-2-quinolylhydrazone complex.[89] Many metals, including Co, Cu, Fe, and Ni were masked by dimethylglyoxime and ammonium citrate (the latter was primarily to prevent precipitation as a result of hydrolysis) when Zn was determined at pH 8 as its 1-(2-thiazolylazo)-2 naphthol complex.[90] If Hg(II) or Mn(II) was present, it could be masked by thiosulfate or periodate ion, respectively.[90]

Other examples include the use of EDTA and cyanide ion when Be is determined with eriochrome cyanine R, and when Bi is extracted as its diethyldithiocarbamate complex. Ammonium citrate and cyanide ion are suitable masking agents when Pb is extracted with dithizone.

The combinations of masking agents that are used depend on the particular kinds of metal ions that are present. Thus, in the determination of Ni with 4-*t*-butyl-1,2-cyclohexanedionedioxime at pH 7, by extraction with xylene, a tiron/ammonium acetate mixture masked Cr(III), Mn(II), Th(IV), Zr(IV), and (after reduction with hydroxylamine) Ce and Sn. When other metals were present they were masked as follows: Cu, with ammonium thiocyanate/sulfite ion; Fe(III), sodium fluoride; rare earths, Ti(IV), and V(V), tartaric acid.[91]

In a spectrophotometric method for Mg at pH 10.5,[3] cyanide ion masks Ag, Cd, Co, Cu, Ni, and Pd (but is inadequate for Fe and Zn), fluoride ion masks Se(IV) and Te(IV) (so long as only small amounts of Ca are present), and triethanolamine masks Al, Fe(III), and In. When Al was determined in steel by extraction of its 8-hydroxyquinolinate into chloroform, at pH 9, cyanide, tartrate, and hydrogen peroxide masked Cd, Cr, Cu, Fe, Mn, Mo, Ni, Sn, Ti, U, V, and Zr.[92] Likewise, cyanide (to mask Cu), tartrate (to prevent precipitation), and EDTA (to mask Pb) were used when Bi was extracted as its diethyldithiocarbamate complex into chloroform, from an aqueous ammonia solution.[93] The combination of cyanide, citrate and EDTA masked Cu, Fe, Mn, and (in part) Pb, when Sb and Sn were determined in ferrous materials by extraction with sodium pyrrolidinedithiocarbamate and chloroform:[94] they were also used, at pH 6.8, to mask Ba, Bi, Cd, Cu, Fe, Ga, Hg, Sr, Ta, V, and Zn when Co was determined spectrophotometrically with PAR.[95]

Mixtures have been proposed for general masking purposes. One example comprised ammonium citrate, ammonium oxalate, sodium thiosulfate, sodium fluoride, and ammonium thiocyanate, for use in the spectrophotometric determination of Ni and Co with pyridyl-2-azo-chromotropic acid in aqueous solution at pH 7.5.[96] This selection masked Al(III), Cd(II), Cu(II), Fe(III), Ga(III), Mn(II), Sn(IV), Ti(IV), Zn(II), and Zr(II).

A combination of ammonium citrate and thiourea prevented the precipitation of Ba(II), Bi(III), Hg(II), Pb(II), Sb(III), and Sr(II).[96]

Another composite buffer and masking solution was made from EDTA, potassium cyanide, sodium fluoride, and sodium and potassium phosphates, for use in the determination of Sb(III) with bromopyrogallol red at pH 6.6–6.8.[97] If necessary, higher-valent ions, and anions such as permanganate or chromate, are reduced to lower oxidation states by addition of hydroxylamine or ascorbic acid prior to masking. When bromopyrogallol red was used to determine Nb(V) at pH 6.0, the solution contained 0.01*M* EDTA, tartrate, and acetate (masking, among others, Mo, Sb, Ta, Ti, and W), and one or more of the following, if needed: fluoride ion (to mask Al and Th), phosphate ion (for U(VI) and Zr), cyanide ion (for Ag), and ascorbic acid (to reduce higher-valent Ce and V).[98]

In the determination of Th with acid alizarine black SN at pH 4.2, mixtures of masking agents were used to eliminate interferences as follows: potassium cyanide (Fe, in the presence of ascorbic acid, Co, Cu and Ni), thioglycolic acid (Bi), potassium iodide (Sb(III) and Sn(IV)), triethanolamine (Ti), sulfosalicylic acid (Al), tiron (W(VI) and Mo(VI)), and ascorbic acid (Ce(IV)).[28] Masking techniques giving Th by difference were needed if U and V were present.[28]

The following masking agents were used when Pd(II) was determined spectrophotometrically with isonitrosoacetylacetone after extraction into carbon tetrachloride from aqueous solution at pH 5–6: mercuric nitrate (to mask sulfite, cyanide and chloride ions), ammonium molybdate (for oxalate), tartaric acid (tungstate), sodium fluoride (Zr, V(V)), potassium sulfate (Cr(III)), ferric chloride (phosphate), sodium molybdate with hydrogen peroxide and sulfuric acid (thiosulfate).[99] In a modification of this method, using benzene extraction from 0.1–1*M* acetic acid, sulfide and sulfite ions were masked by potassium permanganate and sulfuric acid, and EDTA was masked by copper sulfate.[100] A method for the determination of Pt with 3,4-diaminobenzoic acid at pH 10–12 used iodide ion (to mask Hg(II) and Au(III)), EDTA (for Fe(II)), tartrate (for Ni(II), Co(II), Mn(II), Cu(II) and Cd(II)) and formate (for Cr(III)).[101]

All of the above examples have been concerned with the determination of metal ions. No difference in principle exists in the spectrophotometric determination of anions. Thus, when nitrate was determined with chromotropic acid in sulfuric acid, chlorine, and other oxidizing agents were masked by prior reaction with sodium sulfite, nitrite was masked by urea, and chloride ion was masked by adding Sb(III).[102] The latter also demasked any chloro–Fe(III) complexes that were present, thereby discharging their yellow color.[102] In an alternative method for nitrate ion, based on reaction with 2,6-xylenol, sulfamic acid masked nitrite ion and addition of mercuric sulfate masked chloride ion.[103]

REFERENCES

1. T. M. Florence, *Anal. Chem.* **37,** 704 (1965).
2. L. M. Budanova, and S. N. Pinaeva, *Zh. Analit. Khim.* **20,** 320 (1965).
3. R. M. Dagnall, R. Smith, and T. S. West, *Analyst* **92,** 20 (1967).
4. K. L. Cheng, and B. L. Goydish, *Talanta* **13,** 1210 (1966).
5. Y. Shijo, and T. Takeuchi, *Bunseki Kagaku* **14,** 511 (1965).
6. E. G. Owens, and J. H. Yoe, *Anal. Chem.* **32,** 1345 (1960).
7. C. L. Luke, and M. E. Campbell, *Anal. Chem.* **24,** 1056 (1952).
8. H. V. Meek, and C. V. Banks, *Anal. Chem.* **22,** 1512 (1950).
9. M. Otomo, *Bull. Chem. Soc. Japan* **38,** 730 (1965).
10. L. M. Budanova, and S. N. Pinaeva, *Zavodskaya Lab.* **32,** 401 (1966).
11. D. T. Pritchard, *Analyst* **92,** 103 (1967).
12. V. N. Tikhonov, *Zh. Analit Khim.* **20,** 941 (1965).
13. V. Patrovsky, *Talanta* **10,** 175 (1963).
14. B. Budesinsky, and B. Menclova, *Talanta* **14,** 523 (1967).
15. R. J. Romagnoli, and P. F. Lott, *Chemist-Analyst* **52,** 106 (1963).
16. A. W. Ashbrook, and K. Conn, *Chemist-Analyst* **50,** 47 (1961).
17. O. Budevsky, and L. Johnova, *Talanta* **12,** 291 (1965).
18. O. Budevsky, and R. Pribil, *Talanta* **11,** 1313 (1964).
19. V. Patrovsky, *Talanta* **12,** 971 (1965).
20. R. Belcher, T. V. Ramakrishna, and T. S. West, *Chem. Ind.* (*London*) **1963,** 531; *Talanta* **9,** 943 (1962); **10,** 1013 (1963).
21. I. P. Alimarin, and S. B. Savvin, *Talanta* **13,** 689 (1966).
22. E. Lassner, *Chemist-Analyst* **51,** 14 (1962).
23. K. Goto, T. Komatsu, and T. Furukawa, *Anal. Chim. Acta* **27,** 331 (1962).
24. F. Vydra, and R. Pribil, *Chemist-Analyst* **51,** 76 (1962).
25. A. Waksmundzki, and S. Przeczlakowski, *Chem. Anal., Warsaw* **9,** 69 (1964).
26. C. F. Bell, and D. R. Rose, *Talanta* **12,** 696 (1965).
27. M. Hosain, and T. S. West, *Anal. Chim. Acta* **33,** 164 (1965).
28. P. Kusakul, and T. S. West, *Anal. Chim. Acta* **32,** 301 (1965).
29. R. M. Dagnall, and T. S. West, *Talanta* **11,** 1533 (1964).
30. V. Vasak, and V. Sedivec, *Chem. Listy* **45,** 10 (1951); *Coll. Czech. Chem. Comm.* **15,** 1076 (1950).
31. V G. Tiptsova, A. M. Andreichuk, and L. A. Bazhanova, *Zh. Analit. Khim.* **20,** 1200 (1965).
32. T. T. Bykhovtseva, and I. A. Tserkovnitskaya, *Zavodskaya Lab.* **30,** 943 (1964).
33. K. L. Cheng, *Anal. Chem.* **28,** 1738 (1956).
34. K. Toei, and K. Ito, *Talanta* **12,** 773 (1965).
35. G. W. F. Brady, and J. R. Gwilt, *J. Appl. Chem.* **12,** 79 (1962).

36. A. Darbey, *Anal. Chem.* **24**, 373 (1952).
37. R. M. Rush, and J. H. Yoe, *Anal. Chem.* **26**, 1345 (1954).
38. R. E. Mosher, P. J. Burcar, and A. J. Doyle, *Anal. Chem.* **35**, 403 (1963).
39. S. B. Savvin, *Talanta* **8**, 673 (1961).
40. H. N. Ray, S. K. Ray, and M. M. Chakrabartty, *Chemist-Analyst* **55**, 42 (1966).
40a. F. Feigl, *Spot Tests in Inorganic Analysis,* Elsevier, Amsterdam, 1958, p. 171.
41. M. Knizek, and M. Musilova, *Chemist-Analyst* **55**, 108 (1966).
42. G. V. Johnson, and R. A. Young, *Anal. Chem.* **40**, 354 (1968).
43. A. Galik, and M. Knizek, *Talanta* **13**, 1169 (1966).
44. G. Norwitz, and H. Gordon, *Anal. Chem.* **37**, 417 (1965).
45. M. Knizek, and V. Pecenkova, *Zh. Analit. Khim.* **21**, 260 (1966).
46. B. Nebesar, *Anal. Chem.* **36**, 1961 (1964).
47. M. Yosida, and N. Kitamura, *Bunseki Kagaku* **11**, 744 (1962).
48. J. O. Hibbits, H. F. Davis, and M. R. Menke, *Talanta* **8**, 163 (1961).
49. D. F. Boltz, and E. J. Havlena, *Anal. Chim. Acta* **30**, 565 (1964).
50. C. L. Luke, *Anal. Chim. Acta* **37**, 97 (1967).
51. M. Murano, and S. Mujazaki, *Bunseki Kagaku* **15**, 657 (1966).
52. J. Horak, and A. Okac, *Coll. Czech. Chem. Comm.* **28**, 2563 (1963).
53. V. Svoboda, and V. Chromy, *Talanta* **13**, 237 (1966).
54. H. Diehl, and E. B. Buchanan, *Talanta* **1**, 76 (1958).
55. G. Lemoine, and J. Miel, *Ann. Chim. Anal.* **28**, 147 (1946).
56. S. Hara, *Bunseki Kagaku* **14**, 42 (1965).
57. R. M. Dagnall, T. S. West, and P. Young, *Talanta* **12**, 583 (1965).
58. O. A. Ohlweiler, and J. O. Meditsch, *Anal. Chem.* **30**, 450 (1958).
59. S. B. Savvin, *Dokl. Akad. Nauk SSSR* **127**, 1231 (1959).
60. M. Otomo, *Bull. Chem. Soc. Japan* **36**, 137 (1963).
61. F. B. Martinez, and M. P. Castro, *Chemist-Analyst* **48**, 2 (1959).
62. M. Needleman, *Anal. Chem.* **38**, 915 (1966).
63. F. Verbeek, *Bull. Soc. Chem. Belg.* **70**, 423 (1961).
64. B. Budeshinsky, *Zh. Analit. Khim.* **18**, 1071 (1963).
65. U. Tandon, S. N. Tandon, and S. S. Katiyar, *Talanta* **12**, 639 (1965).
66. D. C. Canada, *Anal. Chem.* **39**, 381 (1967).
67. J. J. Lothe, *Anal. Chem.* **28**, 949 (1956).
68. J. P. McKaveney, *Anal. Chem.* **40**, 1276 (1968).
69. K. L. Cheng, and B. L. Goydish, *Anal. Chim. Acta* **30**, 243 (1958).
70. A. E. Vasilevskaya, V. P. Shcherbakov, and A. V. Levchenko, *Zh. Analit. Khim.* **18**, 811 (1963).
71. A. Hulanicki, and M. Minczewska, *Talanta* **14**, 677 (1967).
72. A. Galik, *Talanta* **14**, 731 (1967).
73. E. J. Serfass, W. S. Levine, P. J. Prang, and M. H. Perry, *Plating* **36**, 818 (1949).
74. F. M. Tulyupa, G. E. Bekleshova, and M. A. Vitkina, *Zh. Analit. Khim.* **21**, 783 (1966).
75. D. K. Banerjee, *Anal. Chem.* **29**, 55 (1957).

76. B. K. Afghan, and D. E. Ryan, *Anal. Chim. Acta* **41,** 167 (1968).
77. H. Onishi, and Y. Toita, *Japan Analyst* **14,** 462 (1965).
78. C. Heitner-Wirguin, and M. Gancz, *Talanta* **14,** 671 (1967).
79. N. Kurmaiah, D. Satyanaranya, and V. P. G. Rao, *Talanta* **14,** 495 (1967).
80. R. I. Alekseev, *Zavodskaya Lab.* **7,** 415 (1938).
81. V. F. Mal'tsev, E. N. Pashchenko, and N. P. Volkova, *Zh. Analit. Khim.* **21,** 1205 (1966).
82. T. V. Gurkina, and A. M. Igoshin, *Zh. Analit. Khim.* **20,** 778 (1965).
82a. A. K. Babko, and N. V. Romanova, *Zavod. Lab.* **34,** 1435 (1968).
83. K. L. Cheng, *Anal. Chim. Acta* **28,** 41 (1963).
84. R. Pueschel, and E. Lassner, *Mikrochim. Acta* **1965,** 17.
85. S. S. Yamamura, and J. H. Sikes, *Anal. Chem.* **38,** 793 (1966).
86. V. Patrovsky, *Chem. Listy* **57,** 268 (1963).
87. E. N. Pollock, and L. F. White, *Chemist-Analyst* **52,** 70 (1963).
88. P. Pakalns, and W. W. Flynn, *Analyst* **90,** 300 (1965).
89. S. P. Singhal, and D. E. Ryan, *Anal. Chim. Acta* **37,** 91 (1967).
90. A. Kawase, *Talanta* **12,** 195 (1965).
91. M. M. Barling, and C. V. Banks, *Anal. Chem.* **36,** 2359 (1964).
92. J. L. Kassner, and M. A. Ozier, *Anal. Chem.* **23,** 1453 (1951).
93. J. Kinnunen, and B. Wennerstrand, *Chemist-Analyst* **43,** 88 (1954).
94. E. Kovacs, and H. Guyer, *Z. Anal. Chem.* **208,** 255 (1965).
95. Y. Shijo, and T. Takeuchi, *Japan Analyst* **13,** 536 (1964).
96. A. K. Majumdar, and A. B. Chatterjee, *Talanta* **13,** 821 (1966).
97. D. H. Christopher, and T. S. West, *Talanta* **13,** 507 (1966).
98. R. Belcher, T. V. Ramakrishna, and T. S. West, *Talanta* **12,** 681 (1965).
99. U. B. Talwar, and B. C. Haldar, *Anal. Chem.* **38,** 1929 (1966).
100. U. B. Talwar, and B. C. Haldar, *Anal. Chim. Acta* **39,** 264 (1967).
101. L. D. Johnson, and G. H. Ayres, *Anal. Chem.* **38,** 1218 (1966).
102. P. W. West, and T. P. Ramachandran, *Anal. Chim. Acta* **35,** 317 (1966).
103. D. W. W. Andrews, *Analyst* **89,** 730 (1964).

CHAPTER

8

ION EXCHANGE

A solid ion-exchange resin is essentially an insoluble matrix to which are covalently attached anionic or cationic groups. Electrical neutrality is achieved by the adsorption of ions of opposite charge which are free to exchange with other ions in the solution. Tervalent ions are usually more strongly adsorbed by commercial ion-exchange resins than are divalent ions and these, in turn, are held more strongly than univalent ones, but within any one of these groups the differences between ions are sometimes rather small. Nevertheless, by using appropriate techniques, ion exchange is often a convenient method for the separation of ionic species, either from one another or from neutral molecules. It is commonly used in the fractionation of ions having similar analytical properties and in the removal of interfering ions. The principles involved in ion-exchange processes, and applications to analytical chemistry, have been fully discussed elsewhere.[1-4]

For the separation of mixtures of cations two different techniques are used. In the first of these all cations are adsorbed onto an ion-exchange column, followed by washing and selective elution: An extension of this method has been the development of selective chelating resins containing particular groups such as cyclohexanediaminetetraacetate, or saturated with complexing anions such as EDTA or citrate. The second approach uses selective masking agents which are added prior to perfusion through the column, so that certain of the cations are masked against adsorption and pass through while other cations are retained on the column. It is only the latter of these two techniques which falls within the scope of the present review.

The sign of the charge carried by a species is the most important factor in determining whether, or to what extent, the species will be retained during passage through an ion-exchange column. This provides the basis of almost all methods of ion-exchange separation that depend upon masking. For convenience these methods can be divided into three groups, depending on whether a metal ion is masked by oxidation to form a stable anion, or by complex formation with either an inorganic or an organic ligand.

MASKING OF METAL IONS BY OXIDATION TO ANIONS

Several metal ions can be readily oxidized to stable and soluble anions. Examples include Cr, Mn, Mo, Re, V, and W, giving chromate, permanganate, molybdate, perrhenate, vanadate, and tungstate, respectively, and also heteropolyanions such as phosphomolybdate, phosphotungstate, and silicotungstate. The resulting anions are readily separated from any cations by ion exchange. As an example, V can be masked against adsorption on to a cation-exchange column, and hence be separated from alkali metals. Conversely, Cr, Mo, and V can be separated from iron and other constituents of steel by adsorption as their anions on to an anion-exchange resin.[5] Permanganate is exceptional in that it is a strong oxidizing agent, attacking cation-exchange resins and thereby undergoing reduction, so that this is not a suitable method for separations involving Mn.

Of much wider application, however, is the separation of metal ions based on anionic complex formation with suitable ligands.

MASKING OF METAL IONS BY INORGANIC LIGANDS

In strong solutions of chloride ion, almost all metals form chloro complexes. This provides a basis for the quantitative separation of metal ions, using plots of the logarithm of their distribution coefficient on a strongly basic anion exchanger against strength of hydrochloric acid. Such plots are available for almost all elements of the periodic table;[6] for example, radionuclides have been separated in biological samples, using an anion-exchange resin and hydrochloric, hydrofluoric, and sulfuric acids.[7] Polyvalent ions such as Zr(IV), Hf(IV), Nb(V), Ta(V), and Pa(V) give tailing on such columns because of polymerization, unless hydrofluoric acid is also present,[8] presumably to diminish further the concentrations of the free metal ions.

Recently cation-exchange distribution coefficients on a sulfonated polystyrene resin have been published for forty-five elements in hydrochloric acid/ethanol media, and possibilities of separations have been demonstrated.[8a]

The extent to which complex formation occurs will determine whether a metal will be present as a cation, a neutral complex, or an anionic species. This makes it possible to achieve differences in separations on cation-exchange resins, and appropriate conditions can be selected, using published[9] distribution coefficients for forty-three cations on such a column that have been obtained over a range of hydrochloric acid concentrations. Thus in 3–4*M* hydrochloric acid Zr(IV) is not adsorbed by a cation-exchange resin, whereas Th(IV) is.[10] In concentrated hydrochloric acid containing ascorbic acid as a reducing agent U(IV) is present as an anion, whereas Pu(III) is a cation, so that U(IV) can be removed by adsorption on an anion exchanger.[11] Masking of Fe(III) by conversion to anionic chloro-complexes has been used to separate Fe(III) from Mo(VI), W(VI) and U(VI) by passage of solutions in hydrochloric acid through an anion-exchange resin.[12] The same procedure was used in isolating traces of Th(IV) in biological samples following coprecipitation with ferric hydroxide, but in this case the Th(IV) was retained by passage through a cation-exchange resin.[13] In 0.1*M* hydrochloric acid, Hg(II) is masked against uptake by a cation-exchange resin, and hence can be separated from Cd, Cu, Fe, Ni, Ti, and Zn.[14] Passage of an ammonium chloride solution containing Pd and Ir through a cation-exchange resin leads to retention of Pd as $Pd(NH_3)_4^{2+}$, whereas Ir passes through as $IrCl_6^{2-}$ or $IrCl_6^{3-}$.[15]

Anion-exchange selectivity data are available for hydrofluoric acid solutions of upward of 50 elements,[8,16] simplifying the choice of suitable masking conditions for many other separations; for example, traces of rare earths in Zr can be isolated on a cation-exchange resin by converting Zr to fluozirconate ion.[17] Masking Fe as its anionic complex in dilute hydrofluoric acid enables it to be removed in the same way from Ag, Co, Cu, Mg, Mn, Ni, Pb, and Zn.[18] In 1*M* hydrofluoric acid, Pb, Co, Cu(II), Mn(II), Ni(II), and Cr(III) (partly) are adsorbed on to a cation-exchange resin (H^+ form), whereas Ti(IV), Mo(VI), V(IV), and Fe(III) pass through as anions.[19]

Mixtures of halogen acids are also sometimes used. Thus, Ti(IV) and V(IV) can be separated from each other in 0.1*M* hydrochloric acid, 1*M* hydrofluoric acid mixture because Ti(IV) is present mainly as anionic complexes whereas V(IV) is present as a cation. Passage through an anion-exchange column leads to retention of Ti(IV) on the column, whereas with a cation-exchange column it is the V(IV) that is retained.[8]

Other systems where quantitative information is available include nitric acid,[20,21,22] nitric acid–hydrofluoric acid,[23] sulfuric acid,[24,25] and thiocyanate ion–hydrochloric acid.[26]

Applications of dilute sulfate solutions in such separations include the adsorption of Th(IV) and U(VI) on to an anion exchanger,[27] and also Nb, Pa, Zr, and Ru.[28] Further examples of separations involving inorganic anions cover a wide range. The platinum metals (Pd, Rh, Ru) in 0.5*M* nitric acid pass through a cation-exchange column but Bi, Cd, Cu, Hg, and Pb are retained.[29] Rhenium(VII) is adsorbed on to an anion-exchange resin from aqueous methanol containing nitric acid, whereas Mo(VI) is not adsorbed, hence can be separated from it.[30] A mixture of Cd(II) and Zn(II) can be resolved into its components by passing a solution containing excess of iodide ion through a cation-exchange column, the Cd being present as an anionic complex and being thereby masked against adsorption.[31] The titanium perhydroxy complex is strongly held by cation-exchange resins, so that at pH 5 in the presence of hydrogen peroxide, Ti(IV) can be separated from vanadate, molybdate, and tungstate. On the other hand, in the absence of hydrogen peroxide and in dilute hydrochloric or sulfuric acid Ti(IV) is not adsorbed.

Conversion to their cyano complexes is a convenient way to mask metals such as Co, Fe, Ni, and Zn against adsorption on to cation-exchange resins. The ions $Fe(CN)_6^{3-}$, $Fe(CN)_6^{4-}$, and $Co(CN)_6^{3-}$ can be separated in this way, using a column in the H^+ or NH_4^+ form, whereas, for example, Na and K are retained. However, in separating Ni and Zn as their complex cyanides from alkaline and alkaline earth metal ions, the column must be in the NH_4^+ form because H^+ decomposes the complexes.[33] Alternatively, cyano complexes, such as those of Cu and Fe, can be retained on anion-exchange columns (Cl^- form) as a means of separating them from Ca and Mg.[34] Thiocyanate ion masks Cr(III) by forming an anionic complex which is not adsorbed on a cation-exchange column: this provides a method of separation from Al(III).[35]

MASKING OF METAL IONS BY ORGANIC LIGANDS

Possibilities for ion-exchange separations of mixtures of cations are considerably increased in the presence of anionic ligands. The most commonly used complexing agents are EDTA, di- and tricarboxylic acids, singly or in combination. In metal–ion separations in EDTA media pH is the most important variable. Depending on pH, cation or anion exchangers can be used. Thus in alkaline solution Ca and Mg form EDTA complexes which are retained on an anion-exchange column, whereas alkali metals pass through.[36] In neutral or weakly acid solutions, on the other hand, the Ca– and Mg–EDTA complexes are almost completely dissociated so that Ca and Mg are retained on a cation-exchange column, whereas most cations that would interfere in their determination are present as EDTA complexes and pass through without adsorption. Conversely, if an anion-exchange column was used, Ca and Mg would pass through but the anionic complexes would be held.

At pH 1.5–2, Fe(III) forms a complex with EDTA, but Cr(III) does not. Passage through a cation-exchange resin leaves Fe(III) in the effluent and Cr(III) on the column. With an anion-exchanger the position is reversed.[37] Alternatively, in the determination of Al, Cr, and Fe, fusion with an oxidizing agent forms dichromate ion, and titration with EDTA gives Fe(III)–EDTA: passage through a cation-exchange resin then leads to the retention of Al on the column.[38]

By using EDTA at pH 3.5, Al and Fe can be masked against adsorption on to the sodium form of a cation-exchange resin, and thereby separated from Be.[39] A similar method has been described for Be in clays.[40] Traces of Co in Ni and Ni salts can be isolated by adsorption of the positively charged Co(III)–ethylenediamine complex onto a silica gel column at pH 9, in the presence of EDTA which masks Cu, Ni, Tl, and Zn against adsorption.[41] In the same way, the tris phenanthroline–Fe(II) complex is adsorbed onto silica gel, whereas the EDTA and citrate complexes of other metals are anionic and hence not adsorbed: this is the basis of a method for traces of Fe in Ni and W.[42]

Other aminopolycarboxylic acids have been used in similar ways. Over the pH range 7.5–8.5, Ca forms a stable anionic chelate complex with EGTA (ethyleneglycol bis(2-aminoethylether)tetraacetic acid), but Mg is present as Mg^{2+} and can be adsorbed on to a cation exchanger, thereby permitting their separation.[43] At pH 4.0–4.4, CDTA masks Mn but not Ca or Mg against adsorption on to a cation-exchange resin.[44]

Citric acid is also commonly used. At pH 1, molybdate and tungstate give stable complexes with citric acid, so that they can be separated from Cr, Cu, Fe, Pb, Ni, and V(IV), all of which, under these conditions, are retained by a cation-exchange resin.[45] At pH 2.7–3, Zr is taken up from ammonium citrate solution by an anion-exchange resin, but Sr, Pb, Cd, and Cs pass through.[46] The anion-exchange properties of citrate complexes of Co(II), Zn(II), U(IV), and Th(IV) have been discussed.[47] Ammonium tartrate makes possible the separation of Cu from Pb,[48] and Sn from Sb.[49] The rare earths have been selectively sorbed on to a cation-exchange resin from ammonium acetate/acetic acid buffer (pH 4.6) to separate them from Sc.[50]

A combined EDTA, citrate masking solution at pH 5 permits Sr, but not Ca or Mg, to be retained on a cation-exchange column.[51] On the other hand, if a citrate buffer of pH 3.5–4 is used, Ca and Mg are retained on such a column and can be eluted subsequently with alkaline EDTA, but Fe and Al are masked and pass through.[52] Anion-exchange studies on the alkaline earths in citrate solutions have shown that, by exploiting differences in the extent of absorption, the alkaline earths can be separated from one another, from the alkali metals, and from the rare earths.[53]

If the complexing anion possesses a high charge, and is present in the solution in excess, it may compete so effectively for sites on an anion-exchange resin that metals present as anionic complexes may not be adsorbed. This difficulty can be avoided by pretreating the anion exchange resin with the masking agent, such as citric acid or EDTA, so that subsequent addition to the solution is not necessary: the pretreated exchanger acts as a chelating resin which retains certain cations while the others remain in the external solution. Use has been made of a citrate-saturated anion-exchange resin to separate Fe^{3+}, VO^{2+}, Co^{2+}, Ni^{2+}, and Cu^{2+} (which form complexes) from the alkali metals.[54] Likewise, the oxalate form of an ion-exchange resin retained Ce, Nb, Pr, Y, and Zr while permitting Cs to pass.[55] The same chemical processes are involved as in, eg, the masking of Zr, Hf, Ti, Fe, and Al, but not Mg, with oxalate ion, against retention on a cation-exchange column, a method proposed for determining Mg in Zr metal.[56] However, the use of ligand-saturated resins falls outside the definition of masking and will not be discussed further.

QUANTITATIVE ASPECTS

Although the foregoing discussion has been qualitative, it is possible to treat ion-exchange equilibria quantitatively. For most purposes, the suitability of possible methods for ion-exchange separations, based on batch treatment or the use of ion-exchange columns and including masking, can be assessed so long as the stability constants are known for the complexes that are formed, together with the ion-exchange constants or the distribution coefficients for metal ions between the resin and the aqueous phase. In general, effects of complex formation are of such overriding importance that factors such as changes in activity coefficient values can be ignored.

The underlying principles have been clearly enunciated and applied to several representative examples.[4] These include the following:

1. The use of HEDTA (2-hydroxyethylethylenediaminetriacetic acid) to mask Cu(II) in a mixture of Ca(II) and Cu(II) so that only Ca(II) is adsorbed on to a cation exchanger in a batch extraction;
2. The masking of Fe(III) with EDTA, while permitting La(III) to be adsorbed quantitatively;
3. The separation of Cu(II) and Pb(II), by batchwise operation of a cation-exchange resin, using EDTA and tetren, in which Cu(II) exists predominantly as its positively charged tetren complex, whereas Pb(II) is present as $PbEDTA^{2-}$;
4. The separation of Ca and Ba as their CDTA complexes;
5. The separation of Sr from Ca by cation-exchange, column filtration of their EDTA (or, better, EGTA) solutions, Sr being retained;

6. The separation of Mn, Ca, and Ba by column filtration in the presence of CDTA;

7. The separation of Ca and Mg by column filtration in the presence of EGTA;

8. The separation of Fe(III) and Ti(IV) on an anion-exchange column in the presence of chloride ion.[4]

RESIN SPOT TESTS

A special application of masking in relation to ion-exchange resins is in spot testing, where the resin is used as a means of effecting a significant increase in the concentration of a wanted species. A spot test for Cd(II) is based on the adsorption of CdI_4^{2-} onto a resin bead, followed by reaction with glyoxal bis(2-hydroxyanil). Specificity is claimed if, at the time of CdI_4^{2-} is adsorbed, the solution contains, as masking agents, tartrate (to prevent hydroxide formation), fluoride (to mask Fe(III), U(VI), Ca, Sr and Ba), and thiosulfate (to mask Cu(II) and Au(III)).[57]

Similarly, Co is detected by a blue color on an anion-exchange resin in neutral or slightly acid solution if thiocyanate ion is present. Interference by Fe(III) is masked by sodium fluoride.[58] Conversely, Ti(IV) gives, with hydrogen peroxide in acid solution, a yellow complex which is adsorbed on to a cation-exchange resin. Interference by fluoride ion is partly masked by beryllium chloride.[58] Resin spot tests and their applications have been reviewed by Fujimoto.[58]

REFERENCES

1. O. Samuelson, *Ion-Exchange Separations in Analytical Chemistry,* Almqvist and Wiksell, Stockholm; Wiley, New York, 1963.
2. R. Kunin, *Ion-Exchange Resins,* 2nd ed., Wiley, New York, 1958.
3. F. Helfferich, *Ionenaustauscher,* Verlag Chemie, Weinheim, 1959.
4. A. Ringbom, *Complexation in Analytical Chemistry,* Wiley, New York, 1963, Chap. 6.
5. F. M. Hall, and A. Bryson, *Anal. Chim. Acta* **24,** 138 (1961).
6. K. A. Kraus, and F. Nelson, *Proc. Intern. Conf. Peaceful Uses Atomic Energy, Geneva* **7,** 113 (1956).
7. A. L. Boni, *Anal. Chem.* **32,** 599 (1960).
8. F. Nelson, R. M. Rush, and K. A. Kraus, *J. Am. Chem. Soc.* **82,** 339 (1960).

8a. F. W. E. Strelow, C. R. Van Zyl, and C. J. C. Bothma, *Anal. Chim. Acta* **45,** 81 (1969).

9. F. W. E. Strelow, *Anal. Chem.* **32,** 1185 (1960); see also C. K. Mann, and C. L. Swanson, *Anal. Chem.* **33,** 459 (1961).
10. I. P. Alimarin, and A. M. Medvedeva, *Zh. Analit. Khim.* **22,** 436 (1967).

11. J. A. Hayden, *Talanta* **14,** 721 (1967).
12. K. A. Kraus, F. Nelson, and G. E. Moore, *J. Am. Chem. Soc.* **77,** 3972 (1955).
13. T. Y. Toribara, and L. Koval, *Talanta* **14,** 403 (1967).
14. I. Ubaldini, and S. Cassata, *Ann. Chim.* (*Rome*) **48,** 205 (1958).
15. W. M. MacNevin, and W. B. Crummett, *Anal. Chim. Acta* **10,** 323 (1954).
16. J. P. Faris, *Anal. Chem.* **32,** 520 (1960).
17. H. J. Hettel, and V. A. Fassel, *Anal. Chem.* **27,** 1311 (1955).
18. L. Danielsson, and T. Ekström, *Acta Chem. Scand.* **20,** 2402, 2415 (1966).
19. A. G. Hamza, and J. B. Headridge, *Analyst* **91,** 237 (1966).
20. J. P. Faris, and R. F. Buchanan, *Anal. Chem.* **36,** 1157 (1964).
21. J. P. Faris, and R. F. Buchanan, *U.S. At. Energy Comm. Rept. TID-7606* (1961).
22. F. Ishikawa, S. Uruno, and H. Imai, *Bull. Chem. Soc. Japan* **34,** 952 (1961).
23. E. A. Huff, *Anal. Chem.* **36,** 1921 (1964).
24. F. W. E. Strelow, and C. J. C. Bothma, *Anal. Chem.* **39,** 595 (1967).
25. L. Danielsson, *Acta Chem. Scand.* **19,** 670 (1965).
26. K. Kawabuchi, *J. Chromatog.* **17,** 567 (1965).
27. K. A. Kraus, and F. Nelson, *ASTM Spec. Techn. Publ.* **195,** 27 (1958).
28. L. R. Bunney, N. E. Ballou, J. Pascual, and S. Foti, *Anal. Chem.* **31,** 324 (1959).
29. B. K. Preobrazhenskii, and Y. P. Saikov, *Radiokhimiya* **2,** 68 (1960).
30. J. Korkisch, and F. Feik, *Anal. Chim. Acta* **37,** 364 (1967).
31. S. Kallmann, H. Oberthin, and R. Liu, *Anal. Chem.* **32,** 58 (1960).
32. O. Samuelson, *Svensk Kem. Tidskr.* **57,** 158 (1945).
33. S. Yokusuka, *Bunseki Kagaku* **5,** 74 (1956).
34. J. W. McCoy, *Anal. Chim. Acta* **6,** 259 (1952).
35. N. P. Panchev, and B. Evtimova, *Compt. Rend. Acad. Bulg. Sci.* **18,** 1127 (1965).
36. O. Samuelson, and E. Sjöström, *Anal. Chem.* **26,** 1908 (1954).
37. G. N. Babachev, and L. A. Raeva-Nikolova, *Khim. Ind.* (*Sofia*) **38,** 12 (1966).
38. G. N. Babachev, *Zh. Analit. Khim.* **21,** 881 (1966).
39. M. N. Nadkarni, M. S. Varde, and V. T. Athavale, *Anal. Chim. Acta* **16,** 421 (1957).
40. J. R. Merrill, M. Honda, and J. R. Arnold, *Anal. Chem.* **32,** 1420 (1960).
41. F. Vydra, and V. Markova, *Chemist-Analyst* **54,** 69 (1965).
42. F. Vydra, and M. Kopanica, *Chemist-Analyst* **52,** 88 (1963).
43. P. Povondra, and R. Pribil, *Talanta* **10,** 713 (1963).
44. P. Povondra, and R. Pribil, *Chemist-Analyst* **49,** 109 (1960).
45. R. Klement, *Z. Anal. Chem.* **136,** 17 (1952).
46. S. M. Khopkar, and A. K. De, *Anal. Chim. Acta* **24,** 134 (1961).
47. N. C. Li, and J. M. White, *J. Inorg. Nucl. Chem.* **16,** 131 (1960).
48. A. P. Kreshkov, and E. N. Sayushkina, *Trudy Mosk. Khim. Tekhnol. Inst. im. D.I. Mendeleeva* **1956,** (22), 116.
49. Z. Marczenko, *Chem. Anal.* (*Warsaw*) **2,** 255 (1957).

50. J. Yoshimura, Y. Takashima, and H. Waki, *J. Chem. Soc. Japan, Pure Chem. Sect.* **79,** 1169 (1958).
51. R. D. Ibbett, *Analyst* **92,** 417 (1967).
52. P. Povondra, and Z. Sulcek, *Coll. Czech. Chem. Comm.* **24,** 2398 (1959).
53. F. Nelson, and K. A. Kraus, *J. Am. Chem. Soc.* **77,** 801 (1955).
54. O. Samuelson, and K. Schramm, *Z. Elektrochem.* **57,** 207 (1953).
55. S. Yajima, E. Shikata, and C. Yamaguchi, *Bunseki Kagaku* **7,** 721 (1958).
56. S. V. Elinson, and M. S. Limonik, *Zavodskaya Lab.* **24,** 1434 (1957).
57. P. W. West, and J. Diffee, *Anal. Chim. Acta* **25,** 399 (1961).
58. M. Fujimoto, *Chemist-Analyst* **49,** 4 (1960).

CHAPTER

9

SOLVENT EXTRACTION

There are two main advantages in the use of solvent extraction in analytical chemistry to transfer a metal complex from one solvent, usually water, to another. It provides a convenient method by which to obtain separations from interfering species, and it often permits a useful concentration of the complex to be achieved. These considerations are important in spectrophotometric methods of analysis as well as in purification techniques.

Factors governing the extraction of metal complexes into organic solvents are well understood and have been reviewed elsewhere.[1,2,3] In many cases sufficient quantitative information exists for new methods to be devised from tabulated extraction constants or values of $pH_{1/2}$ (the pH at which the metal ion is distributed equally between aqueous and organic phases) under specified experimental conditions.[3]

The extent of separation from other species can frequently be improved by the use of masking agents such as aminopolycarboxylic, citric, and tartaric acids which form stable, water soluble, nonextractable complexes with the potentially interfering cations.

A quantitative approach to the treatment of masking in solvent extractions has been described by Ringbom[4] and Stary,[3] and can be outlined by considering an example where a metal ion, M^{n+}, reacts with a charged ligand, A^-, to form a neutral complex, MA_n, which is the only species extracted into the organic phase. Because A^- can also accept a proton to form the free acid, HA, the extent of complex formation, and hence of extraction, is pH-dependent. The curve for percentage of metal extracted versus pH is sigmoid, its steepness depending on the charge on the central metal ion. Thus, an increase of one unit in the pH increases the distribution ratio tenfold for univalent cations, but 100-, 1000- and 10,000-fold, respectively, for di-, ter- and tetravalent ions. It is often possible to remove unwanted ions by making a preliminary extraction under conditions where the desired species remains in the aqueous phase but some or all of the interfering species pass into the organic phase. This may be achieved by using the same reagent but at a different pH, or another kind of reagent can be introduced, e.g., thiocyanate to remove most of the Co(II) from a solution before extracting Ni(II) as its dimethylglyoximate.

In extraction systems the position is further complicated by other reactions such as hydrolysis, the formation of other complexes in the aqueous phase, and distribution of the reagent between the two phases. Account can be taken of these by introducing α coefficients as described in Chapter **2**, to calculate conditional extraction constants, and the treatment can readily be extended to include masking.

In the above system, suppose a masking agent, H_xB, to be added in excess, to form nonextractable complexes with the metal. The distribution ratio of metal between organic and aqueous phases becomes

$$q' = \frac{[MA_n]_{org}}{[M] + [MB] + [MB_2] + \cdots}$$

$$= \frac{K_{ext}[A]^n_{aq}}{1 + K_1[B] + \beta_2[B]^2 + \cdots}$$

where K_{ext} is the extraction constant and K_1, β_2, . . . are the stability constants of MB, MB_2, ... Because of protonation equilibria, [B] is pH-dependent and is given by

$$[B] = \frac{[B]_{total}}{1 + [H^+]/K_{a1} + [H^+]^2/K_{a1}K_{a2} + [H^+]^3/K_{a1}K_{a2}K_{a3} + \cdots}$$

These relationships enable the quantitative calculation of the effectiveness of H_xB as a masking agent. Qualitatively they show its dependence on the stability constants of its complexes and on the pK_a values of the free acid, decreasing rapidly at pH values less than the numerically greatest pK_a.

Worked examples using such an approach are available for the following:

1. The masking by EDTA of the extraction of zinc dithizonate into carbon tetrachloride and a comparison with other masking agents;[4]
2. The masking by thiosemicarbazide so as to secure the extraction of copper from its dithizonate in carbon tetrachloride back into water at pH 1.5;[4]
3. The masking by EDTA of the extraction of copper 8-hydroxyquinolinate into chloroform.[3]

The pH region for satisfactory extraction of a metal complex is determined in part by the partition coefficient, P_{HA}, of the neutral molecule HA between the organic and aqueous phases, and by the value of $(pK_a)_{HA}$. In general, the higher the value of $(pK_a + \log p_{HA})$ the more alkaline must be the solution.[3] The solubility of metal complexes in organic and aqueous phases also serves as a guide to the suitability or otherwise of particular systems for solvent extraction. Best results are obtained where the metal complex is almost insoluble in water but is readily soluble in the organic phase.

Again, because organic solvents can usually extract only neutral molecules or ion pairs, the most satisfactory masking agents are those that form anionic complexes with the metal ions that they mask, particularly if, in these complexes, there are many hydrophilic groups to enhance a preference for aqueous rather than nonaqueous solutions. Some examples of masking in systems involving solvent extraction have already been given in Chapter 7. The present discussion will be limited to a qualitative review of applications of masking in solvent extraction, considered in terms of factors such as the nature of bonding groups and the stereochemistry of the metal ion.

Almost all examples where masking is important in solvent extraction concern systems where metal chelates are extracted. However, some exceptions should be noted: They belong to the following types:

1. Formation of simple salts with lipophilic reagents. Rare earths are extracted by butyric acid/chloroform at pH 4–5.5. Sulfosalicylic acid can be used to mask Al, Fe, Nb, Sn, Ta, Ti, and Zr.[5]

2. Formation of ion pairs. In the extraction and spectrophotometric determination of Nb with tetraphenylarsonium chloride, Nb is masked initially against formation and extraction of its thiocyanate complex by adding fluoride ion to a solution 3–5*M* in hydrochloric acid. These conditions permit the extraction of Fe(II), Mo(V,VI), and W(V,VI). Addition of boric acid demasks Nb so that the insoluble ion-association product, tetraphenylarsonium–thiocyanato–niobate, is extracted into chloroform/acetone.[6]

3. Masking of anions. The extraction of UO_2^{2+} into trioctylamine is incomplete if fluoride ion is present, but boric acid or Al(III), added in excess, masks the fluoride ion so that extraction proceeds normally.[7]

9.1 ORGANIC REAGENTS BONDING THROUGH OXYGEN

β-DIKETONES

Acetylacetone, $CH_3COCH_2COCH_3$, readily loses one of its enolizable protons ($pK_a = 8.9$) to give a monoanion which is able to react with many metal ions to form neutral chelates. These can be extracted into excess acetylacetone, either alone or in admixture with organic solvents such as chloroform, carbon tetrachloride, or benzene. Because bonding is through the two oxygen atoms, "hard" metal ions form the most stable complexes, and stability constants increase with the valence of the metal ion. Diminution in the number of coordinated water molecules is also important. With octahedral symmetry, a tervalent cation requires three acetylacetonate ions to neutralize its charge and to occupy the six coordination sites surrounding it to give a complex of predominantly nonpolar character. With a divalent ion, on the other hand, only two acetylacetonate ions are required and, if the structure of the complex approximates to octahedral symmetry, two coordination sites on the metal remain occupied by water molecules so that the solubility of the complex in water is improved and it is less readily extracted. Examples include Co(II), Mn(II), Pb(II), and Zn(II). With tetrahedral or square-planar divalent ions the coordination number is only four, so that no water molecules are involved and extractability is improved. These considerations explain why acetylacetone is a good extractant for Be(II), Cu(II), and Pd(II), as well as for ions such as Fe(III), In(III), Sc(III), Th(IV), and U(VI). The extractability of metal ions into acetylacetone in benzene follows the sequence:[3] Pd(II) > Tl(III) > Fe(III) > Pu(IV) > Be(II) > U(IV) > Ga(III) > Cu(II) > Sc(III) > Al(III) > In(III) > U(VI) > Th(IV) > Pb(II) > Ni(II) > La(III) > Co(II) > Zn(II) > Mn(II) > Mg(II).

The Be– and U(VI)–EDTA complexes have low stability constants, so that EDTA is a good masking agent for most metal ions when Be[8,9] and U(VI)[10] are extracted using acetylacetone. More usually, dibenzoylmethane, $C_6H_5COCH_2COC_6H_5$, is the preferred extractant for U(VI), because of its better solubility in benzene, but the principle is the same and EDTA[11] or cyclohexanediaminetetraacetic acid[12] is again used as a general masking agent.

Reduction with ascorbic acid, together with the addition of cyanide ion, prevent interference from Cu(II), Ti(IV), and Pd(II) when In is extracted as its acetylacetonate into toluene.[13] Advantage can be taken of the inertness of Cr(III) towards complex formation to extract other metals first before heating under reflux to form the Cr(III) complex.[14]

By appropriate substitution in the acetylacetone molecule its pK_a can be lowered and its solubility can be modified. Trifluoroacetylacetone (pK_a 6.3) and thenoyltrifluoroacetone (pK_a 6.2) can be used over a lower pH range than acetylacetone, while the thiophen moiety of thenoyltrifluoroacetone and the benzene rings of benzoylacetone and dibenzoylmethane decrease water solubility but enhance solution in benzene.

Trifluoroacetylacetone extracts Be(II) at pH 5–7 into benzene, with EDTA masking Co, Fe, and Zn.[15] Similarly, EDTA masks Co, Cu, and Ni when Pd(II) and Zn(II) are extracted at pH 8–10 into chloroform, using trifluoroacetylacetone and isobutylamine.[16] Picolinic acid has also been used to mask Cu and Ni, in the extraction of Al (at pH 7.5–9.5) and Fe(III) (pH 2.5) into benzene.[17]

Metal complexes of 2-thenoyltrifluoroacetone are much more readily extracted into benzene and other solvents of low dielectric constant than are the corresponding acetylacetone complexes. In some cases the complexes are also sufficiently colored to be of use in spectrophotometric determinations. Rhenium has been estimated in this way after extraction into an isoamyl alcohol–benzene mixture from 7–9*N* sulfuric acid, while V(V) and Ti(IV) were retained in the aqueous phase by forming peroxy complexes with hydrogen peroxide.[18] Hydrogen peroxide similarly masked Nb and Pa in the extraction of Zr(IV).[19] In the same way, extraction from 5–9*M* hydrochloric acid afforded Pt,[20] while Pd was extracted over the pH range 4.5–8.8, with sodium fluoride to mask Fe(III) and tartaric acid to mask Co, Mn, and Ni.[20]

Other ligands binding through oxygen atoms include cupferron, the tropolones, and the hydroxamic acids. The metals they extract, and masking agents suitable for use with them, are similar to those discussed above; for example, cupferron quantitatively extracts Ti(IV) from a solution containing EDTA as a general masking agent.[21] In the extraction of Zr(IV), oxalic acid masks W.[22]

ORGANIC PHOSPHORUS COMPOUNDS

Organic phosphorus compounds such as tributyl phosphate that are used as extractants, frequently dissolved in an inert solvent as diluent, are immiscible with water but are able to displace water molecules surrounding neutral metal complexes to form species which can be extracted into the organic phase. Coordination through the oxygen of the $P \rightarrow O$ group is probably involved. By this means Pu, U, and Zr can be extracted from nitric acid solution into 30% tributyl phosphate in kerosene, separation from one another being achieved by suitable selection of masking agents.[23] The separation of U from Zr is improved in the presence of arsenazo, pyrogallol, tiron, or oxalic acid, all of which complex strongly with Zr(IV).[24]

In the tributyl phosphate extraction of Ga(III) from 4*M* hydrochloric acid interference by Fe(III) is eliminated by reduction to Fe(II) with ascorbic acid, W(VI) is masked by tartaric acid, and Ag(I) by persulfate oxidation.[25] In a similar extraction of Tl(III) from 2*M* hydrochloric acid Fe(III) is masked by fluoride ion, and Sn(II) is oxidized by bromine water to Sn(IV).[25] If the acidity is increased to 8–10*M* hydrochloric acid, tartaric acid is no longer able to mask W(VI) because ionization of the acid is inhibited. It can still mask vanadium, but this is by reduction of V(V) to V(IV). Hydrogen peroxide masks Ti(IV).[26] Extraction of Nb(V) from strongly acid solutions in the presence of masking agents such as oxalic acid, citric acid, EDTA, nitrilotriacetic acid, tiron, pyrogallolsulfonic acid, and disodium 1,8-dihydroxynaphthalene-3,6-disulfonate has also been studied.[27]

Masking by sodium fluoride retained Ce, Np(IV), Th, and Zr in solution when uranium was extracted from 2*M* nitric acid, using trioctylphosphine oxide. Sulfamic acid kept Fe reduced to Fe(II), and Pu to Pu(IV), so that neither was extracted.[28]

Closely related to masking is the technique of stripping, by which species not wanted in the organic phase are removed by back-extraction with an aqueous solution which either contains appropriate complex forming species or has been adjusted to a suitable pH at which the distribution of the unwanted species strongly favors the aqueous phase. Oxalate prevents the extraction of Bi and Sb by alkylphosphoric acids in heptane from aqueous sulfuric, nitric or perchloric acid, but does not affect In, Sn(II), or Sn(IV), so that oxalate can be used either as a masking agent or as a stripping agent to remove Bi and Sb from such extracts.[29] Similarly, the masking of Ga, In, Sb(III), Sn(II), and Sn(IV) by halide ions makes it possible to strip these cations from the organic phase by shaking with dilute hydrochloric acid, into which they pass as anionic complexes.[29]

9.2 ORGANIC REAGENTS BONDING THROUGH NITROGEN

Anionic ligands in which the metal ion is bonded only through nitrogen atoms are rare, the only familiar examples being dimethylglyoxime and its derivatives. Even in these cases, anions are formed only because one of the hydroxyl hydrogens ionizes, and the extractability of the Ni(II) and Pd(II) complexes into organic solvents is a consequence of the strong internal hydrogen bonding of the —OH and $—O^-$ groups in the planar *bis* complexes. This effect does not operate for the complexes of other metals, because they are not planar, so that these groups are strongly solvated and greatly diminish extraction into organic solvents. Hence dimethylglyoxime and its derivatives such as α-furildioxime show high selectivity, and the main application of masking is to prevent hydrolytic precipitation of ter- and tetravalent metals in the pH range (7–12) that is used in the extraction. Ammonium tartrate or citrate is suitable for this purpose.[30,31] Nickel is extracted into chloroform as its dimethylglyoxime complex in the presence of Co if the latter is first masked by cyanide ion and the $Ni(CN)_4^{2-}$ is carefully demasked by hydrogen peroxide or formaldehyde. Cobalt can also be masked as $Co(NH_3)_6^{3+}$. Copper and manganese can be masked by reduction with thiosulfate[32] and hydroxylamine hydrochloride,[33] respectively. Thioglycolic acid masked Cu(II) when Ni(II) was extracted into toluene with 4-methyl-1,2-cyclohexanedionedioxime.[34] Masking procedures when the isopropyl derivative was used included thioacetamide for Cu(II) and fluoride ion for Fe.[35]

Ion-pair formation provides the other main examples of extractions where only metal–nitrogen bonding is important. Even where the ligand is a neutral species, so that its metal complex is a cation, extraction of the resulting complex is often possible if a suitable ion of opposite charge is present in the system to form an ion association product. An indirect determination of Hg(II) is based on the extraction into dichloroethane of the ion pair formed by HgI_4^{2-} and the 2,2′-bipyridine-Fe(II) chelate in a solution of pH 6.3, followed by measurement of the amount of Fe(II) chelate in the organic phase. EDTA is used as a general masking agent.[36] Iron can be determined in samples of Co by extraction of the 1,10-phenanthroline–Fe(II) iodide into chloroform, after Co(II) is oxidized to Co(III) and masked with EDTA.[37] A similar determination of Fe in U is possible if U is masked by tartrate at pH 6–7, and of Fe in Cu if the latter is masked by mercaptoacetate at pH 9.[38] The estimation of Cu as its Cu(I) chloro complex with neocuproine (2,9-dimethyl-1,10-phenanthroline) can be made specific if the complex is extracted at pH 4–6 into chloroform in the presence of EDTA and citrate as masking agents.[39,40]

9.3 ORGANIC REAGENTS BONDING THROUGH SULFUR

The dithiocarbamates are not very selective in their complexing behavior. They react with most metals that are able to form insoluble sulfides, to give colored complexes that are usually insoluble in water but soluble in suitable organic solvents. Careful choice of masking conditions is usually necessary if these reagents are used for analytical purposes. The stability sequence for these complexes lies in the following approximate order: Hg > Ag > Cu > Ni > Co(II) > Bi > Pb > Cd > Tl(III) > Sb(III) > Zn > Mn(II) > Fe(III).

One possible way of improving the selectivity of analyses with dithiocarbamates is to use metal-exchange reactions based on this sequence. An example would be copper diethyldithiocarbamate as a reagent for Hg, in accordance with the reaction[41]:

$$Hg^{2+} + Cu(DDC)_2 \rightarrow Cu^{2+} + Hg(DDC)_2$$

Metal ions lying to the right in the above sequence cannot displace Cu in this way.

Another technique that can be used is chelate solvent extraction, in which the organic phase contains the chelating agent, such as diethylammonium diethyldithiocarbamate or zinc dibenzyldithiocarbamate, whereas the aqueous phase constitutes a solution of metal ions, acid or base for pH adjustment, and appropriate masking agents. One of the advantages of this technique is that it can be employed over wider pH ranges and at higher ligand concentrations than might otherwise be possible. If extraction from acid solution is required, diethylammonium diethyldithiocarbamate is more suitable, because it is more stable when used in this way, than is sodium diethyldithiocarbamate in the aqueous phase.[42,43]

EDTA and citric acid are satisfactory as masking agents in the extraction of Cu diethyldithiocarbamate into butyl acetate when Cu is determined in steel.[44] In such a mixture, EDTA masks Ni and Pb, while a citrate buffer (pH 5.5) prevents precipitation of Al.[45] EDTA is also a good general masking agent when Ag or As is extracted.

In the determination of radioactive Co by benzene extraction as its diethyldithiocarbamate from water at pH 5–5.5 complexes of the radionuclides Mn, Cu, and Fe were also extracted but were removed by washing the extract with nonradioactive mercuric chloride solution. The Mn, Cu, and Fe complexes were dissociated, with the formation of the Hg complex, and the metal ions passed into the aqueous phase, but Co, present as Co(III) complex, persisted unchanged.[46]

Cyanide ion effectively masks Cu, Au, and Hg, so that cyanide ion and EDTA, or cyanide ion and tartrate, have been used as masking agents in the extraction of Bi[47] and Cd,[48] respectively, into carbon tetrachloride as their diethyldithiocarbamates. The use of 1-(2-pyridylazo)-2-naphthol makes it possible to mask Zn without affecting the extraction of Cd.[49]

Further applications of masking agents to improve the possibilities of analytical separations with dithiocarbamates have recently been discussed.[41]

Diethylammonium diethyldithiocarbamate extracts the same metals as sodium diethyldithiocarbamate, but it can be used in very acid solutions. Under these conditions Cu can be masked by the addition of potassium iodide,[50] or it can be separated from Bi, Pb, and other metals by back-extraction from the organic phase with cyanide ion.

Other sulfur-bonding chelates include the xanthates and dithiols, such as toluene-3,4-dithiol. The latter has been used for the determination of Mo in tungsten ores and other materials, using citric acid as a masking agent for tungsten in the strongly acid solutions.[51]

9.4 ORGANIC REAGENTS BONDING THROUGH OXYGEN AND NITROGEN

The commonest examples in this group comprise 8-hydroxyquinoline, nitrosophenol, and their derivatives. In all cases the complex-forming ability is strongly pH-dependent, but neutral complexes are formed with a wide range of metal ions, so that selectivity requires good pH control and the use of masking agents.

8-Hydroxyquinoline complexes with polyvalent species, such as Al(III), Fe(III), Tl(III), V(V), Mo(VI), and U(VI), and a small number of divalent ions, including Mg, Ca, Sr, Ba, Cu, Co, and Ni, are extractable into solvents such as chloroform, the sequence being the following[3]:

$$\begin{aligned} &\mathrm{Pd(II)} > \mathrm{Mo(VI)} > \mathrm{W(VI)} > \mathrm{V(V)} > \mathrm{Tl(III)} > \mathrm{Fe(III)} \\ &\quad > \mathrm{Zr(IV)} > \mathrm{Ga(III)} > \mathrm{Cu(II)} > \mathrm{Ti(IV)} > \mathrm{In(III)} > \mathrm{Bi(III)} \\ &\quad > \mathrm{Ni(II)} > \mathrm{U(VI)} > \mathrm{Al(III)} > \mathrm{Th(IV)} > \mathrm{Co(II)} > \mathrm{Zn(II)} \\ &\quad > \mathrm{Sc(III)} > \mathrm{Cd(II)} > \mathrm{Pb(II)} > \mathrm{Mn(II)} > \mathrm{Be(II)} > \mathrm{La(III)} \\ &\qquad > \mathrm{Ag(I)} > \mathrm{Mg(II)} > \mathrm{Ca(II)} > \mathrm{Sr(II)} > \mathrm{Ba(II)}. \end{aligned}$$

In addition, and for reasons similar to those discussed under acetylacetone, 8-hydroxyquinoline complexes with a number of divalent cations are formed in water but are only poorly extracted.

Interference by Cd, Co, Cu, Ni, and Zn in the presence of Al can be masked by $0.3M$ cyanide ion.[52] Iron(III) can be reduced to Fe(II) and converted to ferrocyanide.[53] At pH 4.5, Zr(IV) is masked by quinalizarinsulfonic acid.[54] At pH 8.5–9, when Fe(III) is reduced by sulfite ion to Fe(II) and EDTA and cyanide ion are used jointly as masking agents, not only Al but also Ti, V, Ta, Nb, U, Bi, Ga, In, Sb, and Zr are extracted. The first five of these can be removed by repeating the extraction, at pH 7.5–8.5, in the presence of hydrogen peroxide.[55,56] Saturated ammonium carbonate solution (pH 9.5–10) masks U(VI).[57]

EDTA is also a good masking agent in the extraction of Mo(VI), U(VI), and W.[58] At pH 2–3, if Mo(VI) is reduced by hydrazine to Mo(V), which forms a stable complex with EDTA, it can be prevented from reacting with 8-hydroxyquinoline. Under the same conditions W(VI) is stable and forms a complex with 8-hydroxyquinoline that can be extracted with a butanol/chloroform mixture, so that small amounts of tungsten can be determined in molybdenum and its compounds.[59] The converse is also possible: molybdenum can be determined in the presence of tungsten by extracting its colored complex with toluene-3,4-dithiol into amyl acetate or chloroform in the presence of citric acid to mask tungsten.[60]

With Ca–EDTA as masking agent, Cu can be selectively extracted with 8-hydroxyquinoline in the presence of Co, Ni, and Fe(III),[53] whereas ordinarily EDTA masks Cu. Other masking procedures include the use of tartrate to mask Ca, Sr, and Ba when Mg was extracted with 8-hydroxyquinoline as a tetra-*n*-butylammonium ion pair,[61] the reduction of Fe(III) to Fe(II) in the presence of 1,10-phenanthroline,[62] and the use of 4-sulfobenzenearsonic acid or 6*M* ammonium acetate to mask Th.[62]

8-Hydroxyquinaldine (2-methyl-8-hydroxyquinoline) is similar to 8-hydroxyquinoline in most of its properties, except that is does not extract Al(III). Cyanide ion is suitable for masking Cd, Fe, Ni, and Zn when Be(II) is extracted at pH 7.5–8.5 into chloroform.[63,64]

1-Nitroso-2-naphthol forms a *tris* complex with Co(III) which, because of its low-spin state, is appreciably more stable than the Fe(III), Cu(II), or Ni(II) complexes, so that Co(III) can be separated from most metal ions by extraction of its complex into chloroform, benzene, or several other solvents from an aqueous solution at pH 2.5–5. Use of a citrate buffer prevents the precipitation of metal hydroxides and masks Fe(III). The Co complex, once formed, is kinetically inert, so that the organic extract can be washed with acid, alkali, cyanide, or EDTA solutions to remove interfering ions such as Cu(II). When U(VI) is extracted as its 1-nitroso-2-naphtholate into isoamyl alcohol from aqueous solution at pH 4.5–7.5, the presence of EDTA masks V and Fe(III).[65]

The isomeric 2-nitroso-1-naphtol is chemically similar and is also used for extracting Co(III), masking Fe(III) with citrate[66] or fluoride,[67] Sn(II) by oxidation, and Mn by treatment with hydrogen peroxide.[68] EDTA is used as a masking agent in the extraction of Pd(II) into benzene or toluene from a solution at pH 1.0–2.5.[69]

9.5 ORGANIC REAGENTS BONDING THROUGH SULFUR AND NITROGEN

Because the bonding in metal–dithizone complexes is through sulfur and nitrogen, the stability sequence for such complexes lies in a similar order to that for diethyldithiocarbamate complexes, but differences between successive metals are probably not quite as great. The p*K* of dithizone, as an acid, is 4.55 ($\mu = 0.1$) (69a), so that pH control provides one method of regulating the extraction of metal ions. Thus the Hg(II) complex can still be extracted from 6*N* sulfuric acid, whereas for Zn optimum extraction is from near-neutral solutions. Dithizone is usually employed in a chelate solvent extraction technique, using a solution of dithizone in chloroform or carbon tetrachloride. Cations can be masked in aqueous solutions or, alternatively, back-extraction from the organic phase can be used to improve selectivity.[70]

EDTA in dilute acid masks Cu, but allows Ag and Hg to be extracted.[70] At any pH it also masks Bi, Cd, Co, Ni, Pb, Tl(III), and Zn.[71]

If Ca–EDTA, instead of EDTA, is used, many metal ions are still masked at pH 9 but Cu is extracted by dithizone.[71] Thiosulfate is also a useful masking agent. In 0.5*M* sulfuric acid thiosulfate masks Hg but not Cu,[71] while at pH 4–5 it masks most cations except Zn, Sn(II), and Pd (but Pd can be masked by cyanide ion).[72] Alkaline tartrate (pH 9.2–9.9) masks Ga[73] and W,[74] while permitting the extraction by dithizone of Bi, Cd, Co, Cu, Ni, Pb, and Zn. This can be used to remove any of these metals as impurities or to determine their concentrations in samples of Ga and W. Cyanide can also be included in the alkaline tartrate or citrate solution to increase the number of metal ions, including Ni and Co, that are masked. Metal ions that can be extracted under these conditions include Bi, Cd, In, Pb, Sn, and Tl, but if the pH is raised above 12, Bi and In are no longer extracted and above pH 13 Pb also remains behind.

In dilute acid solutions, Ag and Hg are masked by chloride, bromide, or iodide, but Au, Cu, and Pd can be extracted. Thiocyanate ion does not mask Au, Cu, or Hg. Addition of cyanide ion retains Au in the aqueous phase. Oxidation of Pt(II), Sn(II), and Tl(I) to Pt(IV), Sn(IV), and Tl(III), and reduction of Fe(III) with sulfur dioxide to Fe(II) mask these metals against dithizone extraction.

The bonding through sulfur atoms by bis(carboxymethyl)dithiocarbamate, to form water-soluble complexes, makes this ligand a useful one for masking the metal ions, such as Cu, Pb, Cd, and Ni that react readily with dithizone, so that at pH 4.5 only Zn is extracted by dithizone in carbon tetrachloride.[75] Diethanolaminedithiocarbamate behaves similarly and can be used when Zn is extracted from slightly alkaline solutions.[76]

Another reagent belonging to this group is 8-mercaptoquinoline which extracts into chloroform or benzene most metals that form insoluble sulfides. The more important masking agents are alkaline potassium cyanide (masking Ag, Au, Co, Fe, Ir, Ni, Os, Pd, Pt, and Ru), thiourea (Ag, Au, Cu, Hg, Os, Pt, Ru), and sodium fluoride (Fe, Sn). Ions not extracted from concentrated hydrochloric acid include Ag, Bi, Co, Fe, Hg, Mo, and Sn.

9.6 SUBSTOICHIOMETRIC EXTRACTION

In this procedure for the selective determination of traces of metals, a single extraction is made of the metal ion to be determined, in the form of a complex with an organic reagent, the reagent being present in less than the amount needed to complex all of the metal.[77,78] This technique, which avoids the necessity to separate all of the desired metal, can only be used in isotopic dilution or neutron activation analyses. Originally suggested for use with an ion-exchange column, to retain the desired metal as its EDTA complex, it has since been extended to solvent extraction.[78] Application of the technique has been reviewed.[79] Although it uses masking procedures and extractions similar to conventional methods, it is potentially more selective.

In many cases interfering metals are removed by preliminary extractions with suitable complexing agents, but some examples of masking can be given. Silver has been extracted with dithizone, using EDTA as a general masking agent.[80] Manganese was determined by activation analysis, following substoichiometric extraction of Mn by thenoyltrifluoroacetone into ethyl acetate from an aqueous solution at pH 8–8.5, in which Fe, Hf, Sc, Th, and Zr were masked with tartrate and fluoride ions, Cu was masked with ammonia, and Co with cyanide ion.[81] Ammonium tartrate was a suitable masking agent for Zr and Sn, and partly for Ti and W, when Ga was extracted as its 8-hydroxyquinoline complex for determination by neutron activation.[82]

REFERENCES

1. G. H. Morrison, and H. Freiser, *Solvent Extraction in Analytical Chemistry,* Wiley, New York, 1957.
2. D. D. Perrin, *Organic Complexing Reagents,* Interscience, New York, 1964, Chap. 10.
3. J. Stary, *The Solvent Extraction of Metal Chelates,* Pergamon Press, Oxford, 1964.
4. A. Ringbom, *Complexation in Analytical Chemistry,* Interscience, New York, 1963, Chap. 7.
5. L. L. Galkina, and L. A. Glazunova, *Zh. Analit. Khim.* **21,** 1058 (1966).
6. H. E. Affsprung, and J. L. Robinson, *Anal. Chim. Acta* **37,** 81 (1967).
7. D. J. Crouse, and K. B. Brown, *U.S. Atomic Energy Commission Report, ORNL-1959.*
8. J. A. Adam, E. Booth, and J. D. H. Strickland, *Anal. Chim. Acta* **6,** 462 (1952).
9. T. Shigematsu, M. Tabushi, and F. Isojima, *Bunseki Kagaku* **11,** 752 (1962).
10. A. Krishen, and H. Freiser, *Anal. Chem.* **29,** 288 (1957).
11. R. Pribil, and M. Jelinek, *Chem. Listy* **47,** 1326 (1953).
12. J. Stary, and E. Hladky, *Anal. Chim. Acta* **28,** 227 (1963).
13. B. K. Afghan, R. M. Dagnall, and K. C. Thompson, *Talanta* **14,** 715 (1967).
14. J. P. McKaveney, and H. Freiser, *Anal. Chem.* **30,** 1965 (1958).
15. W. G. Scribner, M. J. Borchers, and W. J. Treat, *Anal. Chem.* **38,** 1779 (1966).
16. W. G. Scribner, and A. M. Kotecki, *Anal. Chem.* **37,** 1304 (1965).
17. G. P. Morie, and T. R. Sweet, *Anal. Chim. Acta* **34,** 314 (1966).

18. A. K. De, and M. S. Rahaman, *Talanta* **12,** 343 (1965).
19. F. L. Moore, *Anal. Chem.* **28,** 997 (1956).
20. A. K. De, and M. S. Rahaman, *Analyst* **89,** 795 (1964).
21. K. L. Cheng, *Anal. Chem.* **30,** 1941 (1958).
22. O. I. Popova, and V. I. Kornilova, *Zh. Analit. Khim.* **16,** 651 (1961).
23. M. Beran, K. Zdrazil, and S. Havelka, *Coll. Czech. Chem. Comm.* **30,** 2850 (1965).
24. M. Kyrs, R. Caletka, and P. Selucky, *Coll. Czech. Chem. Comm.* **28,** 3337 (1963).
25. A. K. De, *Talanta* **14,** 629 (1967).
26. A. K. De, and M. S. Rahaman, *Talanta* **11,** 601 (1964).
27. C. Konecny, *Radiochim. Acta* **6,** 192 (1966).
28. R. J. Baltisberger, *Anal. Chem.* **36,** 2369 (1964).
29. I. S. Levin, A. A. Shatalova, T. G. Azarenko, I. A. Vorsina, N. A. Burtovaya-Balakireva, and T. F. Rodina, *Talanta* **14,** 801 (1967).
30. A. Claassen, and L. Bastings, *Rec. Trav. Chim. Pays-Bas* **73,** 783 (1954).
31. W. Nielsch, and L. Giefer, *Mikrochim. Acta* **1956,** 522.
32. W. Nielsch, *Z. Anal. Chem.* **150,** 114 (1956).
33. W. Oelschlager, *Z. Anal. Chem.* **146,** 339, 346 (1955).
34. P. D. Blundy, and M. P. Simpson, *Analyst* **83,** 558 (1958).
35. B. L. McDowell, A. S. Meyer, R. E. Feathers, and J. C. White, *Anal. Chem.* **31,** 931 (1959).
36. K. Kotsuji, *Bull. Chem. Soc. Japan* **38,** 402 (1965).
37. F. Vydra, and R. Pribil, *Talanta* **3,** 72 (1959).
38. F. Vydra, and M. Kopanica, *Chemist-Analyst* **52,** 88 (1963).
39. A. R. Gahler, *Anal. Chem.* **26,** 577 (1954).
40. B. W. Bailey, R. M. Dagnall, and T. S. West, *Talanta* **13,** 753 (1966).
41. A. Hulanicki, *Talanta* **14,** 1371 (1967).
42. K. Lounama, *Z. Anal. Chem.* **146,** 422 (1955).
43. K. Nishimura, and T. Imai, *Bunseki Kagaku* **13,** 713 (1964).
44. J. L. Hague, E. D. Brown, and H. A. Bright, *J. Research Natl. Bur. Standards* **47,** 380 (1951).
45. G. Norwitz, J. Cohen, and M. E. Everett, *Anal. Chem.* **36,** 142 (1964).
46. K. Motojima, S. Bando, and N. Tamura, *Talanta* **14,** 1179 (1967).
47. K. L. Cheng, R. H. Bray, and S. W. Melsted, *Anal. Chem.* **27,** 24 (1955).
48. H. Bode, *Z. Anal. Chem.* **144,** 165 (1955).
49. W. Berger, and H. Elvers, *Z. Anal. Chem.* **171,** 255 (1959).
50. D. Abson, and A. G. Lipscomb, *Analyst* **82,** 152 (1957).
51. P. G. Jeffery, *Analyst* **82,** 558 (1957).
52. C. H. R. Gentry, and L. G. Sherrington, *Analyst* **71,** 432 (1946).
53. C. H. R. Gentry, and L. G. Sherrington, *Analyst* **75,** 17 (1950).
54. J. P. Riley, and H. P. Williams, *Mikrochim. Acta* **1959,** 825.
55. A. Claassen, L. Bastings, and J. Visser, *Anal. Chim. Acta* **10,** 373 (1954).
56. J. L. Kassner, and M. A. Ozier, *Anal. Chem.* **23,** 1453 (1951).
57. A. W. Ashbrook, and G. M. Ritchey, *Can. J. Chem.* **39,** 1109 (1961).
58. H. Freiser, *Chemist-Analyst* **50,** 94 (1961).

59. A. V. Vinogradov, and M. I. Dronova, *Zh. Analit. Khim.* **20,** 343 (1965).
60. C. F. Bickford, W. S. Jones, and J. S. Krene, *J. Am. Pharm. Assoc., Sci. Ed.* **37,** 255 (1948).
61. S. J. Jankowski, and H. Freiser, *Anal. Chem.* **33,** 776 (1961).
62. D. W. Margerum, W. Sprain, and C. V. Banks, *Anal. Chem.* **25,** 249 (1953).
63. K. Motojima, *Bull. Chem. Soc. Japan* **29,** 71 (1956).
64. K. Motojima, and H. Hashitani, *Bunseki Kagaku* **9,** 151 (1960).
65. I. P. Alimarin, and Y. A. Zolotov, *Zh. Analit. Khim.* **12,** 176 (1957).
66. A. Claassen, and A. Daamen, *Anal. Chim. Acta* **12,** 547 (1955).
67. D. Monnier, W. Haerdi, and J. Vogel, *Helv. Chim. Acta* **44,** 897 (1961).
68. H. Schueller, *Mikrochim. Acta* **1959,** 107.
69. L. E. Ross, G. Kesser, and E. T. Lucera, *Anal. Chem.* **32,** 1367 (1960).
69a. R. W. Geiger, and E. B. Sandell, *Anal. Chim. Acta* **8,** 197 (1953).
70. H. Friedeberg, *Anal. Chem.* **27,** 305 (1955).
71. H. Freiser, *Chemist-Analyst* **50,** 62 (1961).
72. H. Fischer, and G. Leopoldi, *Z. Anal. Chem.* **107,** 241 (1936).
73. A. Galik, and M. Knizek, *Talanta* **13,** 589 (1966).
74. G. L. Hubbard, and T. E. Green, *Anal. Chem.* **38,** 428 (1966).
75. A. Hulanicki, and M. Minczewska, *Talanta* **14,** 677 (1967).
76. A. Zeman, J. Ruzicka, and J. Stary, *Talanta* **10,** 685 (1963).
77. J. Ruzicka, and J. Stary, *Talanta* **8,** 228 (1961).
78. J. Ruzicka, J. Stary, and A. Zeman, *Talanta* **11,** 1151 (1964).
79. J. Stary, and J. Ruzicka, *Talanta* **11,** 697 (1964).
80. A. Zeman, J. Ruzicka, and J. Stary, *Talanta* **10,** 905 (1963).
81. F. Kukula, B. Mudrova, and M. Krivanek, *Talanta* **14,** 233 (1967).
82. J. Zeman, J. Ruzicka, and V. Kuvik, *Talanta* **13,** 271 (1966).

CHAPTER

10

ELECTROANALYTICAL CHEMISTRY

10.1 POLAROGRAPHY

The quantitative determination of mixtures of ions by polarographic analysis requires that each kind of ion gives a sufficiently characteristic polarographic wave and that there is an adequate separation between the successive half-wave potentials of the various ions. For two ions present in comparable concentrations the minimum separation must be about 0.4 volts for univalent ions and 0.2 volts for divalent ions if their polarographic waves are not to overlap. When the concentrations of the ions differ appreciably, greater separations between half-wave potentials are needed if step heights are to be evaluated directly. Where separations are less than this, it may be necessary to use graphical methods to obtain step heights.

One of the easiest ways of increasing the separation between successive waves is by complex formation. Provided the equilibria involved in metal complex formation and dissociation are established sufficiently rapidly, this usually leads to the displacement of the half-wave potential, $E_{1/2}$, to lower (more negative) values, both for homogeneous reactions and for reactions in which a solid metal or a metal amalgam is produced. In a homogeneous reaction the amount of the displacement depends on the difference in the apparent stability constants of the oxidized and reduced forms of the corresponding complexes. Similarly, where the metal complex is reduced to free metal, the apparent stability constant of the complex is important: In both cases, the displacement of the polarographic wave is ordinarily pH-dependent. Copper(II) ion in a tartrate medium is a typical example. In acid solutions, the logarithm of the apparent stability constant of the copper(II)–tartrate complex, and hence $E_{1/2}$, varies linearly with pH, but in neutral or alkaline solutions relationships are less simple, probably because hydrolyzed species and species bonding through ionized hydroxyl groups are also formed.[1]

In the polarography of many organic compounds, this dependence of $E_{1/2}$ on pH can often be used, by careful buffering of the solution, to obtain the maximum separation between the waves for different compounds, especially in cases where $E_{1/2}$ for one substance is pH-dependent but $E_{1/2}$ for the other substance is constant. Even among related compounds, such as carboxylic acids, separations can sometimes be achieved by taking advantage of differences in their pK_a values and hence in the shapes of their $E_{1/2}$ —pH curves. Thus at pH 7, in ammonium phosphate buffer, maleic acid can be determined polarographically in the presence of fumaric acid.[1a] Similarly, at pH 9.2 progesterone can be determined in the presence of methyl testosterone.[1b]

When two metal ions form complexes with the same ligand, the difference between the relevant stability constants will determine the extent to which the half-wave potentials are further separated. For present purposes, masking will be taken to include any process by which otherwise overlapping steps are separated sufficiently to permit resolution without the removal of species from the solution. This may be by complex formation, change in valence state, or both, as in the addition of cyanide ion to Cu(II) solution to form cuprocyanide ion. It may be by precipitation without filtration, as in the addition of sulfate ion to precipitate lead as lead sulfate from a lead-zinc mixture. Alternatively, in some cases, it is possible to use electrochemical masking, in which adsorption of surface-active substances on to a dropping mercury electrode prevents complex ions of one or more of the major constituents from reaching the electrode surface.

Many analytically useful $E_{1/2}$ values are available for inorganic ions in supporting electrolytes, with and without complex-forming species,[2] so that a choice can frequently be made of suitable experimental conditions for carrying out polarographic determinations on specified mixtures of metal ions. In cases where existing information does not permit this to be done, it may be necessary to resort to preliminary physical methods of separation, such as electrolytic stripping at controlled cathode potential (commonly by using a mercury cathode[3]), distillation, extraction, precipitation, or passage through a suitable ion-exchange column. Precipitation is frequently unsatisfactory because some of the wanted species becomes adsorbed on the precipitate.

The efficiency of electrolytic separations can be improved by adjustment of pH and by using masking agents. In the presence of tartrate and cyanide ions, lead can be separated from many other metals by deposition onto a platinum cathode from solutions at pH 9–10.[4] Antimony can be similarly separated by using an EDTA solution at pH 8.5–9.5 unless iron(III) is present: in this case the pH has to be 0.5–1.0.[4] Using a mercury cathode, Cd(II) and Pb(II) can be separated from Sn(IV) if the latter is masked by a tartrate solution maintained at pH 9.[5] With appropriate choices of pH and masking agents mixtures containing Ag, Bi, Cd, and Cu can also be resolved.[6]

MASKING BY CHANGE IN VALENCE STATE

Ferric ion gives a well-defined wave due to reduction to Fe(II), with a half-wave potential in 0.5*M* sodium citrate given (in volts) by

$$E_{1/2} = 0.426 - 0.108 \text{ pH}$$

as pH varies from 4 to 12 at 25°. This wave is eliminated if Fe(III), dissolved in dilute nitric acid, is reduced with ascorbic acid, thereby permitting Cu(II) to be determined polarographically in steel.[7] Similarly, reduction of Fe(III) in hydrochloric acid solution with hydroxylamine or hydrazine hydrochloride has been used to mask iron in the determination of Bi, Cd, Cr, Cu, In, Pb, Sb, Sn, Ti, and Sn.[8,9,10] In the polarographic determination of Sn at pH 1–2 in EDTA solutions, As and Sb were masked by prior oxidation with potassium permanganate.[11] Cobalt has been determined in the presence of excess nickel(II) by carefully oxidizing the Co(II) to Co(III) with lead dioxide in an acidic oxalate solution: under these conditions the Co(III) wave precedes that of Ni(II).[12]

MASKING BY COMPLEX FORMATION

The commonest masking agents used in polarography are EDTA, cyanide, and tartrate ions. EDTA is useful when general masking is required. In this way, curves due to Al, Be, Co, Cr, Mn, Ni, Ti, and Zn were eliminated in the determination of U(VI),[13] and those due to Bi, Fe, and Pb when Tl(I) was determined in biological materials and in lead.[14] Below pH 6, $E_{1/2}$ for Tl(I) is almost unaffected by EDTA, unlike most of the di- and trivalent cations. This makes it possible to determine traces of Tl in Cd by pulse polarography, in an EDTA solution at pH 4.3.[15]

Frequently, it is necessary to suppress only one kind of metal ion, and here again EDTA is useful because its complexing ability is strongly pH-dependent. In nitric acid containing ascorbic acid as supporting electrolyte, polarographic curves due to Bi and Cu are superimposed. ($E_{1/2} = 0.0$ volts in 0.1*M* HNO_3), so that only the total height can be measured. Addition of EDTA masks Bi, leaving the curve due to Cu, thereby permitting the determination of both Bi and Cu. This method has been applied to the analysis of impurities in Pb[16] and In.[17] At pH 3.2, on the other hand, EDTA masks Cu(II) but not U(VI), and this property has been applied to the estimation of U(VI) in the presence of Cu(II).[18] Use of EDTA as a masking agent also enables Te to be determined polarographically in alkaline solution in the presence of Pb.[19]

Similarly, addition of EDTA to an ammoniacal solution of Co(II) and Ni(II) displaces the curve for Ni(II) sufficiently that Co(II) can be determined.[20] Cyclohexanediaminetetraacetic acid resembles EDTA in its properties and, likewise, it has been used to mask lead in the determination of Tl(I), polarographically and by amperometric titration with potassium chromate at pH 5.5.[21]

Cyanide ion is a suitable masking agent for Zn and Cu in the polarographic determination of Pb[22,23] and Ni[24] in bronzes, brasses, and other copper-base alloys, with triethanolamine to mask Pb and Ni against precipitation in the alkaline solution. A similar combination of cyanide ion and triethanolamine in alkaline solution has been used in determining traces of Fe(III) in Cu-containing materials.[25] (Lead interferes.) If, instead, the solution is buffered to pH 6–7 with sodium acetate, and 0.1*M* EDTA is present, the wave due to Fe(III) occurs about 0.3 volts before the Cu(II) wave.[25] A mixture of cyanide ion and EDTA in a pH 9.5 ammonia buffer containing 15% ethanol masks Pb and Cu while permitting Bi to be determined ($E_{1/2} = -0.81$ volt).[26]

Most of the above examples are as would be expected from the masking behavior of the particular ligands. In the same way, alkaline tartrate displaces the curve due to Fe(III), permitting the determination of Cu in the presence of excess Fe(III).[27] Fluoride ion masks Fe(III) in a method for Ni.[28] Citrate[9] and pyrophosphate[29] have been used to shift the Fe(III) step to a suitable potential, away from those of Cu and Pb, for determining Fe in Zn alloys. At pH 5.5–6.0 Fe(III) and Ti(IV) can be determined simultaneously in a citrate buffer: their $E_{1/2}$ values differ by more than 0.6 volts.[30] Aluminum in Th compounds has been determined by linear-sweep oscillographic polarography, using the dye 5-sulfo-2′,4′,2-trihydroxyazobenzene and masking Th with acetate at pH 5.8.[31] In alkaline triethanolamine and tartrate solution, the half-wave potentials of Cu and Fe are adequately separated (—0.48 and —0.96 volts) for Cu and Fe to be determined in Al alloys, the tartrate also masking Al against precipitation.[32]

Complex formation between organic compounds and buffer components can also sometimes be used for masking. Thus waves due to compounds containing the groups

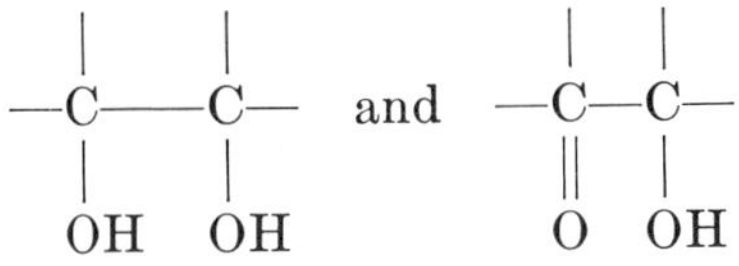

can be diminished or completely suppressed by adding borate buffers. Examples include benzil[32a] and ninhydrin.[32b] Similarly, the polarographic waves given by carbonyl compounds are displaced in the presence of primary amines.

ELECTROCHEMICAL MASKING

Surface-active materials have a marked effect on electrode processes, and can lead, in some cases, to the slowing down or inhibition of a particular reaction as a result of the absorption of the surface-active material at the electrode. For example, gelatin and other surface-active agents often have striking influences on polarograms, oscillographic polarograms, chronopotentiograms, and other electrical processes, leading to shifts of half-wave potentials, decreases in limiting currents, the appearance of minima, the splitting of waves, and even their elimination. The polarographic wave of the EDTA-Bi(III) ion can be completely obliterated by small amounts of gelatin.[33]

More importantly for the present discussion, one or more such waves may be displaced or eliminated without affecting other polarographic waves in a particular system. All such cases are examples of *Electrochemical Masking,* a concept due originally to Reilley and his coworkers.[33,34] The influence of surface-active materials on electrode reactions has been attributed directly to their adsorption on the electrode surface,[35–39] and other important properties include the size of the adsorbed molecule and its electronic charge.

In the presence of complexing species, the diffusion currents of metal ions are not always additive. When a mixture of bismuth and lead in a medium containing EDTA is subjected to polarographic analysis, the Bi–EDTA wave, which appears first, is normal and obeys the Ilkovic equation. The wave produced by Pb is not independent of the concentration of Bi. This is because EDTA liberated in the reduction of Bi–EDTA diffuses back from the mercury surface and results in a diminished wave for Pb. If Triton X-100 is added, the reduction wave of Bi–EDTA is masked and the wave for Pb becomes linear with Pb concentration.[34]

Some of the earliest examples of electrochemical masking include the action of camphor and gelatin on the reduction of cystine,[40] camphor on chloroplatinate,[41] gelatin on Cu(II) in tartrate,[42] camphor on Mo(VI) in tartrate,[43] and quinoline on Cu(II) in citrate.[44] In the polarographic determination of Tl(I) in sodium triphosphate buffer, the addition of 0.1% camphor displaces the cathodic steps of Pb(II), Cu(II), Fe(III), and Bi(III), but does not affect Tl.[45] Similar behavior is shown by Ag(I) in pyrophosphate–triphosphate buffer.[46] The presence of 0.005% Triton X-100 permits the determination of Tl(I) in admixture with a 500-fold excess of Pb.[47] By the combined actions of Triton X-100 and EDTA to mask Cu or Bi, it is possible to determine Tl in Cu,

Pb in Bi, Fe in Cu, and (in the presence of tartrate and isopropanol) Pb in Cu:[48] the Cu–EDTA wave is suppressed by Triton X-100 and camphor, but the Tl is not affected.[49] This is also true of most metal ions when diethylenetriaminepentaacetic acid is used as the complexing agent in pH 4–5 acetate buffer.[50]

Phenols have also been used as surface-active agents. A mixture of 2-naphthol, thymol, and diphenylamine enables Tl, Cd, and In to be determined in trihydroxyglutaric acid at pH 3.[51] When Cd and In are present in large excess, the estimation of traces of Tl can be carried out in 0.1M KCl, pH 3 (HCl) solution containing camphor.[51] Phenol, itself, is suitable in a tartrate buffer (pH 5.4) for determining Cu, Bi, and Sb present in comparable concentration.[51]

Tetrabutylammonium cation strongly inhibits the reduction of most cations but has very little effect on the reduction of complex anions such as Fe(III) or Ti(IV) oxalates.[52] This inhibiting effect also varies with the charge on the cation. Thus in 0.1M perchloric acid tetrabutylammonium sulfate suppresses the wave due to Cd which otherwise overlaps with that due to Tl. To halt the reduction of Tl(I) requires a more powerful reagent such as tribenzylamine with dimethylcetylbenzylammonium chloride. The electrode potential is also important. Uncharged materials such as polyvinyl alcohol, polyethylene glycol *p*-isooctylphenyl ether or the Tritons are active throughout the potential range, but dodecyl sulfonate and other anions inhibit at positive potentials, whereas tetraalkylammonium salts, tribenzylamine, or higher amines are effective at negative potentials.[52]

In some cases, a reaction of the adsorbed ligand with a cation can produce a new wave by which the ligand can be determined: cysteine in concentrations down to $10^{-5}M$ can be determined in this way by reaction with nickel.[52] This technique, in which electrochemical masking is applied to the analysis of surfactant concentrations by using an indicator depolarizer, has also been described by Phillips[53] who determined traces of tetrabutylammonium ion in cadmium chloride solutions using a stationary electrode. He was also able to determine concentrations of cysteine down to $2 \times 10^{-6}M$ by their surfactant effect on the maximum due to oxygen in the polarography of 0.002M KCl solution.[54] Conversely, *o*-phenylenediamine can be estimated in the presence of its isomers because between pH 5.5 and 7.0 it gives a catalytic step at the foot of the wave for reduction of Ni(II) at a dropping mercury electrode.[55] Pyridine behaves similarly.[56]

The effect of the type of surfactant in electrochemical masking is illustrated in the polarographic reduction of U(VI)–Cu(II) mixtures. In sodium sulfate solution dodecyl sulfate masks U(VI) against reduction, whereas Cu(II) is selectively masked by dodecylamine in dilute sulfuric acid, so that Cu(II) and U(VI) can be determined in the presence of each other.[57]

Large amounts of cadmium are electrochemically masked with 0.05% gelatin and substoichiometric amounts of EDTA in ammonia/ammonium chloride buffer, whereas traces of zinc can be determined quantitatively.[58] In triethylenetetraminehexaacetic acid (TTHA), bismuth ordinarily gives a two-step polarographic wave covering the range —0.3 and —1.1 volts, but surface-active agents such as gelatin, tylose, Triton X-45, or X-100 adsorb on to the mercury drops and inhibit one of these steps, so that only the curve with $E_{1/2} = -1.04$ volts is given by Bi-TTHA in 0.1*M* hydrochloric acid. This makes it possible to determine Cu, Pb, Cd, and Bi present as impurities in In.[59]

Adsorbed metal complexes can also exhibit electrochemical masking[60] in a way that is similar to the use of cationic surfactant polymers to achieve selective permission of the electrode reactions of anions.[61] Thus the adsorption of Cu(I) chloride strongly inhibits the reduction of persulfate,[52] and the absorption of lead blocks reduction of Hg(II) from bromide or iodide media.[63] Species reduced at potentials positive to the dissolution of mercury are masked by the absorbed lead; for example, it does not inhibit the reduction of Ag(I) as the bromide or iodide, Au(III) as the bromide, or Fe(III) as the bromide, but it masks Pt(II) as the bromide or iodide. This has been proposed as a method for determining Ag in Ag–Hg mixtures, by measuring step heights with and without added lead.[60]

10.2 AMPEROMETRIC TITRATIONS

Amperometric titrations are related to polarography in that they depend on the use of a dropping mercury, or other micro-, electrode to which is applied a voltage sufficient to produce limiting current conditions, this current, in turn, being proportional to the concentration of the electro-active material. When this material reacts with a suitable reagent, added as titrant, to form a species which is no longer active, the current is diminished. The equivalence point of the titration is obtained graphically from the plot of current versus titre. Considerations governing masking in amperometric titrations are similar to those in polarography, as the following representative examples indicate. They also show the importance of EDTA both as a titrant and as a masking agent.

Fluoride ion masks Al and La in the EDTA titration of Hg(II)[64] and Tl(III).[65] Fluoride ion and tartaric acid also mask Fe(III), Sb(III) and Sn(IV) in the amperometric determination of lead with dimercaptothiopyranones.[65a] Silver nitrate titration of cyanide–thiocyanate mixtures, using a rotating platinum electrode, gives the total concentration of cyanide plus thiocyanate: masking with formaldehyde enables thiocyanate alone to be determined.[66] Chloride ion does not interfere in the amperometric titration of bromide ion with silver nitrate if the solution is $0.01M$ in ammonia. Increasing the ammonia concentration to $0.3M$ masks bromide ion as well, but permits iodide ion to be titrated.[67] In the titration of Ag(I) in ammoniacal solution with sodium diethyldithiocarbamate, EDTA masks Cd, Co, Cu, Fe, Mn, Ni, and Zn, and tartrate masks Pb.[68] Alternatively, Ag(I) can be titrated in ammoniacal solution with iodide ion, using tartrate ion to mask any Fe, Cu, or Pb.[69] Ascorbic acid reduction masks Fe(III) in the determination of Cu(II) in a biamperometric titration with EDTA.[70] At pH 0.9–3.5, Mo(VI) can be titrated with 8-mercaptoquinoline if V(V) is masked by reduction to V(IV), Fe(III) is masked with EDTA, and W(VI) is masked with phosphoric or citric acid.[71] In the EDTA titration of Ca at about pH 12, Mg is masked as $Mg(OH)_2$ or $MgNH_4PO_4$.[72] When ferrocyanide is used to titrate Zn, large amounts of Fe can be masked by adding citric acid before the solution is neutralized with ammonia.[73] Citrate ion has also been used to mask Ni, Co, and Bi in a similar titration for Cd.[74]

10.3 ELECTROGRAVIMETRIC DETERMINATIONS

Electrolytic separations such as those discussed in Section 10.1 can frequently be made quantitative and, in fact, the gravimetric determination of metals by electrodeposition was one of the earliest electroanalytical methods. In many cases, this method can be applied to mixtures of metal ions, so long as adequate differences in deposition potentials can be achieved by the use of complexing agents as supporting electrolytes. Thus, nickel and cobalt can be electrolytically separated from many other metals in the presence of ammonium chloride, tartrate ion, and hydrazine.[75] Addition of cyanide ion, followed by formaldehyde, also masks cobalt whereas nickel is still deposited.[75] Similarly, selenium and tellurium can be separated from a mixture of metal ions if tartrate ion is present as a masking agent.[76] Many other examples of controlled-potential separations and electrogravimetric determinations based on this technique can be devised using published values for deposition potentials.[77]

REFERENCES

1. J. J. Lingane, *Ind. Eng. Chem., Anal. Ed.* **16,** 147 (1944).
1a. P. J. Elving, A. J. Martin, and I. Rosenthal, *Anal. Chem.* **25,** 1082 (1953); P. J. Elving, and I. Rosenthal, *Anal. Chem.* **26,** 1454 (1954).

1b. P. Zuman, J. Tenygl, and M. Brezina, *Coll. Czech. Chem. Comm.* **19,** 46 (1954).
2. See, for example, J. Meites, ed., *Handbook of Analytical Chemistry,* McGraw-Hill Book Co., New York, 1963, Section 5, p. 53.
3. For a simple apparatus of this type, see J. J. Lingane, *Ind. Eng. Chem., Anal. Ed.* **17,** 332 (1945).
4. A. K. Majumdar, and S. G. Bhowal, *Anal. Chim. Acta* **36,** 399 (1966).
5. W. M. Wise, and D. E. Campbell, *Anal. Chem.* **38,** 1079 (1966).
6. A. K. Majumdar, and S. G. Bhowal, *Anal. Chim. Acta* **35,** 206 (1966).
7. M. Kopanica, and S. Vorlicek, *Chemist-Analyst* **54,** 105 (1965).
8. R. Strubl, *Coll. Czech. Chem. Comm.* **10,** 466 (1938).
9. A. S. Nickelson, *Analyst* **71,** 58 (1946).
10. G. H. Osborn, and A. Jewsbury, *Analyst* **73,** 506 (1948).
11. M. Cyrankowska, *Chemia Anal.* **5,** 851 (1960).
12. K. Komarek, *Coll. Czech. Chem. Comm.* **7,** 404 (1935).
13. R. Pribil, and A. Blazek, *Chem. Listy* **45,** 432 (1951); *Coll. Czech. Chem. Chem. Comm.* **16,** 554 (1951).
14. R. Pribil, and Z. Zabransky. *Chem. Listy* **45,** 427 (1951); *Coll. Czech. Chem.* **16,** 554 (1951).
15. E. Temmerman, and F. Verbeek, *J. Electroanal. Chem.* **19,** 423 (1968).
16. J. Musil, and M. Kopanica, *Chemist-Analyst* **55,** 106 (1966).
17. *Ibid.,* p. 9.
18. F. Strafelda, and D. Vondrak, *Coll. Czech. Chem. Comm.* **31,** 4635 (1966).
19. V. V. Sokolov, E. N. Vinogradova, and A. V. Novoselova, *Vest. Mosk. Gos. Univ., Ser. Khim.* **1967,** 50.
20. S. Vicente-Perez, and P. Sanchez Batanero, *Inform. Quim. Anal.* (*Madrid*) **19,** 97 (1965).
21. J. N. Gaur, and D. S. Jain, *Acta Chim. Hung.* **51,** 171 (1967).
22. G. W. C. Milner, *Analyst* **70,** 250 (1945); *Metallurgia* **36,** 287 (1947).
23. M. Spalenka, *Metallwirtschaft* **23,** 341 (1944).
24. G. W. C. Milner, *Analyst* **70,** 468 (1945).
25. P. Souchay, and J. Faucherre, *Anal. Chim. Acta* **3,** 259 (1949).
26. G. Conradi, and M. Kopanica, *Chemist-Analyst* **52,** 11 (1963).
27. W. C. Davies, and C. Key, *Ind. Chem.* **19,** 555 (1943).
28. P. W. West, and J. F. Dean, *Ind. Eng. Chem., Anal. Ed.* **17,** 686 (1945).
29. C. A. Reynolds, and L. B. Rogers, *Anal. Chem.* **21,** 176 (1949).
30. V. Vandenbosch, *Bull. Soc. Chim. Belg.,* **58,** 532 (1949).
31. T. M. Florence, *Anal. Chem.* **34,** 496 (1962).
32. W. Kemula, and S. Rubel, *Chem. Anal.* (*Warsaw*) **10,** 1333 (1965).
32a. R. Pasternak, *Helv. Chim. Acta* **30,** 1948 (1947).
32b. A. K. Vlcek, E. Spalek, and L. Kratky, *Techn. Llidka kozeluzska* **24,** 258 (1949).
33. C. N. Reilley, W. G. Scribner, and C. Temple, *Anal. Chem.* **28,** 450 (1956).
34. R. W. Schmid, and C. N. Reilley, *J. Am. Chem. Soc.* **80,** 2087 (1958).
35. J. Heyrovsky, F. Sorm, and J. Foreijt, *Coll. Czech. Chem. Comm.* **12,** 11 (1947).

36. M. Loshkara, and A. Kryukova, *Dokl. Akad. Nauk SSSR* **62,** 97 (1948).
37. M. Dratovsky, and M. Ebert, *Chem. Listy* **48,** 498 (1954).
38. S. L. Bonting, and B. S. Aussen, *Rec. Trav. Chim. Pays-Bas* **73,** 455 (1954).
39. R. Tamumushi, and T. Yamanaka, *Bull. Chem. Soc. Japan* **28,** 673 (1955).
40. I. M. Kolthoff, and C. Barnum, *J. Am. Chem. Soc.* **63,** 520 (1941).
41. H. A. Laitinen, and E. I. Onstott, *J. Am. Chem. Soc.* **72,** 4565 (1950).
42. L. Meites, *J. Am. Chem. Soc.* **71,** 3269 (1949).
43. E. P. Parry, and M. G. Yakubic, *Anal. Chem.* **26,** 1294 (1954).
44. P. Delahay, and I. Trachtenberg, *J. Am. Chem. Soc.* **79,** 2355 (1957).
45. P. S. Shetty, P. R. Subbaraman, and J. Gupta, *Anal. Chim. Acta* **27,** 429 (1962).
46. P. R. Subbaraman, P. S. Shetty, and J. Gupta, *Anal. Chim. Acta* **26,** 179 (1962).
47. T. Fujinaga, K. Izutsu, and T. Inoue, *Coll. Czech. Chem. Comm.* **30,** 4202 (1965).
48. T. Fujinaga, K. Izutsu, and T. Inoue, *Coll. Czech. Chem. Comm.* **30,** 4209 (1965).
49. T. Fujinaga, and K. Izutsu, *Japan Analyst* **10,** 63 (1961).
50. E. Jacobsen, and G. Kalland, *Anal. Chim. Acta* **29,** 215 (1963).
51. E. N. Vinogradova, G. V. Prokhorova, L. B. Sveshnikova, and L. A. Shorova, *Zh. Analit. Khim.* **21,** 659 (1966).
52. J. Kuta, *Z. Anal. Chem.* **216,** 242 (1966).
53. S. L. Phillips, *Anal. Chem.* **38,** 343 (1966).
54. S. L. Phillips, *Anal. Chem.* **39,** 679 (1967).
55. H. B. Mark, *Anal. Chem.* **36,** 940 (1964).
56. H. B. Mark, and C. N. Reilley, *Anal. Chem.* **35,** 195 (1963).
57. I. M. Issa, *J. Chem. U.A.R.* **8,** 117 (1965).
58. T. Fujinaga, T. Nagai, and T. Inoue, *J. Chem. Soc. Japan, Pure Chem. Sect.* **85,** 110 (1964).
59. G. Conradi, and M. Kopanica, *Chemist-Analyst* **53,** 4 (1964).
60. R. W. Murray, and R. L. McNeely, *Anal. Chem.* **39,** 1661 (1967).
61. Y. F. Frei, and I. R. Miller, *J. Phys. Chem.* **69,** 3018 (1965).
62. I. M. Kolthoff, and R. Woods, *J. Am. Chem. Soc.* **88,** 1371 (1966).
63. D. J. Gross, and R. W. Murray, *Anal. Chem.* **38,** 405 (1966).
64. R. W. Schmid, *Chemist-Analyst* **51,** 56 (1962)
65. F. Vydra, *Talanta* **12,** 139 (1965).
65a. A. K. Arishkevich, and A. N. Alybina, *Zavod. Lab.* **33,** 1367 (1967).
66. S. Musha, and S. Ikeda, *Bunseki Kagaku* **14,** 270 (1965).
67. H. A. Laitinen, W. P. Jennings, and T. D. Parks, *Ind. Eng. Chem., Anal. Ed.* **18,** 355 (1946).
68. V. I. Lotareva, *Zh. Analit. Khim.* **20,** 790 (1965).
69. V. V. Sokolovskii, *Lab. Delo* **8,** (8), 3 (1962).
70. J. Vorlicek, and F. Vydra, *Talanta* **12,** 671 (1965).
71. V. I. Suprunovich, and Y. I. Usatenko, *Zh. Analit. Khim.* **20,** 800 (1965).
72. J. S. Fritz, M. J. Richard, and S. K. Karraka, *Anal. Chem.* **30,** 1347 (1958).

73. A. P. Voiloshnikova, M. T. Kozlovskii, and O. A. Songina, *Zavodskaya Lab.* **23,** 273 (1957).
74. H. C. Saraswat, *Proc. Indian Acad. Sci.* **51A,** 34 (1960).
75. A. K. Majundar, and S. G. Bhowal, *Mikrochim. Acta* **1967,** 1086.
76. A. K. Majundar, and S. G. Bhowal, *Anal. Chim. Acta* **38,** 468 (1967).
77. See, for example, N. Tanaka, *"Electrodeposition,"* in I. M. Kolthoff, and P. J. Elving, eds., *Treatise on Analytical Chemistry.* Interscience, New York, 1963, Part 1, Vol. 4; A. J. Lindsey, *"Electrodeposition,"* in C. L. Wilson, and D. W. Wilson, eds., *Comprehensive Analytical Chemistry,* Elsevier, Amsterdam, 1964, Vol. IIA, Chap. 2.

CHAPTER

11

MASKING OF REACTIONS AND REACTIVITY

11.1 KINETIC MASKING

Measurement of the rate of a catalytic reaction provides a very sensitive method for detecting or determining microamounts of the species responsible for the catalytic effect. Conversely, the procedure can be used for the microdetermination of ions or molecules that are able, under the experimental conditions, to mask the catalyst involved in the reaction. This possibility was pointed out by Yatsimirskii and Fedorova,[1] who coined the term "catalytic titration" or "catalimetric titration" for a procedure which comprises the titration of an inhibitor solution with a solution of a catalyst (or the reverse), using the catalytic activity on the substrate to indicate the equivalence point. They titrated silver nitrate with $10^{-6}M$ solutions of potassium iodide, the catalytic effect being estimated by the oxidation of arsenious acid with cerium(IV), which is catalyzed by iodide ion.[1] Subsequently, a method based on the same reaction was also described for palladium(II).[2]

For such methods to be useful in quantitative analysis, it is essential that the stability constant of the resulting complex is sufficiently large that reactions go almost to completion. Alternatively, the solubility product must be very low. In the original example,[1] the solubility product of silver iodide is 8×10^{-17}, so that, if catalytic activity is detectable at a free iodide concentration of $10^{-8}M$, the equilibrium concentration of silver iodide at this level is also near to $10^{-8}M$: this should make it possible to determine micromolar concentrations of silver ion to within about $\pm$ 1%.

The oxidation of malachite green by periodate ion is catalyzed by manganese(II) ion in dilute acid solutions. Trace amounts of EDTA can be determined by their ability to complex with some of the manganese present, and hence to slow down the reaction.[3] In this type of reaction, a conditional stability constant of at least 10^6 is needed if a sensitivity of $2 \times 10^{-7}M$ is to be achieved for the ligand. This is based on the detectability of a 10% change in a total metal concentration of about $10^{-6}M$. An inherent difficulty with this technique is the lack, in general, of selectivity in such a method, and, in many cases, interference is to be expected from other complex-forming species such as cyanide, iodide, or sulfide ion, polyamines or aminopolycarboxylic acids, and in this particular example, from metal ions that form complexes with EDTA under the experimental conditions. This is clearly shown in a recent method for trace amounts of copper(II)-complexing ligands, which is based on partial masking of the copper(II)-catalyzed aerial oxidation of *L*-ascorbic acid in aqueous pH 6.4 buffer.[4] Substances that can be determined in this way include cysteine, 2-aminoethanethiol, salicylic acid, EDTA, 1,10-phenanthroline, and ethylenediamine.[4] The rate can be adjusted by varying the pH.

Adjustment of pH has been used to improve the selectivity of a method for determining traces of sulfur-containing ligands, based on the inhibition by traces of silver(I) and mercury(II) of the iodide-catalyzed oxidation of arsenious acid by cerium(IV).[5] At pH 1.0, the formation constants of mercury(II) complexes with mercaptoacetic acid, 2-aminoethanethiol, thioacetamide and dithiooxamide are still sufficiently large, whereas those for most complexes with nitrogen or oxygen are not.

By using the metal ion catalyst as a titrant for the solution containing the reaction mixture and the inhibitory species, the concentration of the complexing species can be determined by noting the titer at which the reaction begins to take place. This technique has been applied to determine cyanide concentrations around $10^{-7}M$ by their effect on the copper(II)–ascorbic acid system at pH 10.5.[6]

A method has been described for the detection and determination of ultratrace quantities of metal ions, based on their masking action towards catalysts of coordination chain reactions.[7] The overall reaction

$$\text{Nitrien}^{2+} + \text{CuEDTA}^{2-} \rightarrow \text{Cutrien}^{2-} + \text{NiEDTA}^{2-}$$

which proceeds by the chain-propagating steps

$$\text{EDTA} + \text{Nitrien}^{2+} \rightarrow \text{NiEDTA}^{2-} + \text{trien}$$

$$\text{trien} + \text{CuEDTA}^{2-} \rightarrow \text{Cutrien}^{2+} + \text{EDTA}$$

is catalyzed by traces of the free ligands. Thus, EDTA concentrations down to $10^{-7}M$ are capable of controlling the rate of conversion of $10^{-3}M$ Nitrien^{2+} and CuEDTA^{2-}, while the bright blue color of Cutrien^{2+} provides a convenient basis for spectrophotometric measurements. In this case, addition of $5 \times 10^{-8}M$ of a metal ion forming a stable complex with EDTA would diminish the free EDTA concentration by 50% and hence would halve the reaction rate.[7]

For such a coordination chain reaction to be useful in analysis, several conditions must be fulfilled. The metal ion to be determined must be able to react rapidly with the ligand to form a very stable complex, so that effectively all of the metal present is converted to complex. Chain-propagation, rather than availability of alternative pathways, must account for almost all of the exchange reactions, and free ligand concentrations should remain essentially constant as the reaction proceeds.

A disadvantage is that EDTA forms complexes with many metal ions, but when metal ions are present in low concentrations some selectivity can be achieved by using the effect of pH on metal-complex formation with EDTA as shown by conditional stability constants and the rates of complex formation. At pH 7–8, Fe and Cu can be determined, but Mg is ineffective. At pH 8.9 Mn(II) reacts with EDTA but Cr(III) is kinetically inert toward EDTA.[7] Selectivity can be further improved by including masking agents which do not interfere with the EDTA catalysis. Examples include thiosulfate, to mask Ag(I) and Hg(II), and cyanide ion followed by chloral hydrate, to mask Ni(II), Co(II) and Fe(III).[8]

A method has been proposed for the estimation of low concentrations (0.006–10 μg/ml) of metal ions such as Ag(I), Hg(II), Cu(II), Ni(II), and Zn(II), based on the extent to which they mask cyanide ion and diminish its catalysis of the reduction of *o*-dinitrobenzene by *p*-nitrobenzaldehyde to give a colored product in 0.05*M* sodium hydroxide.[9] Selectivity can be increased by including other masking agents such as EDTA, trien, or fluoride ion in the system.[7]

Traces of vanadium can be determined in water by using its catalysis of the oxidation of gallic acid by persulfate in acid solution. Interference by halides is eliminated by masking with mercuric nitrate, although mercuric ion interferes if present in much excess.[10]

The expression "kinetic masking" has also been applied, with less justification, to reactions that depend on the slowness with which certain complexes are formed or dissociate; for example, in cold solutions chromium(III) reacts sufficiently slowly with EDTA that at pH 6 Co(II), Cu(II), and Ni can be titrated rapidly with EDTA, against naphthylazoxine, in the presence of Cr(III).[10a] A similar procedure has been described for Fe(III) and Ni.[10b] Conversely, the Ni–EDTA complex dissociates only slowly below 5°C, so that Ni can be determined in the presence of other metals such as Cd, Co(II), Pb, Mn(II), Hg(II), Tl(III), and Zn by adding excess EDTA to obtain the total metal content, followed by back titration, at pH 2 and low temperature, with bismuth, against catechol violet.

11.2 MASKING OF CHEMICAL REACTIVITY

In general, oxidation-reduction potentials of complexes formed by transition metal ions differ from the potentials of the corresponding metal ion couples.[11] For example, the equilibrium

$$2\,Fe^{3+} + 2\,I^- \rightleftharpoons 2\,Fe^{2+} + I_2$$

is displaced to the left in the presence of EDTA, tartrate, pyrophosphate, fluoride ion, or other ligands that form more stable complexes with Fe(III) than with Fe(II), whereas, excess iodide ion, an acid solution and the absence of complexing species drives the reaction to the right For a similar reason the sensitivity of Fe(II) to aerial oxidation is greatly enhanced if fluoride ion is present: this effect is eliminated if boric acid is added in excess to mask the fluoride ion.

In the presence of 1,10-phenanthroline, at pH 2–4, Co(II) is oxidized quantitatively by Fe(III) chloride, and can be determined titrimetrically.[12] 2,2′-Bipyridine can be used instead of 1,10-phenanthroline.[13] The addition of 1,10-phenanthroline raises the potential of Fe(II)/Fe(III) sufficiently that U(IV) is rapidly oxidized to U(VI) by Fe(III), at pH 1.5-5 and room temperature.[14]

Just as complex formation can facilitate such reactions, it can also mask others. Thus, in the presence of EDTA, higher oxides of Pb,[15] Mn[15] and Ce[16] are reduced with potassium iodide, but Fe(III) and Cu(II) are masked by the EDTA and do not react. The same procedure can be used for analyzing Fe–Cr and Fe–Mn alloys by oxidizing them to Fe(III), chromate and permanganate.[16]

Examples such as these emphasize the profound effect that complex formation can have on the chemical reactivity of the central metal ion. The reactivity of a ligand is also considerably modified by metal-binding, particularly when chelate ring formation is involved because this brings into close proximity a multivalent cation, with its powerful electron-withdrawing effect while, at the same time, imposing major steric constraints. This can lead to a greatly increased rate of chemical reaction, for example in the metal-catalyzed hydrolysis of amino acid esters and peptides, or even to new ones as in the copper(II) oxidation of cysteine to cystine. However, this topic is outside the scope of the present discussion on the masking of ligand reactivity.

When a ligand takes part in complex formation, many typical reactions are retarded or abolished, particularly if they involve ligand electron pairs that are used in bonding with the metal ion. For example, while ammonia or an amine remains coordinated to a metal ion the amino nitrogen cannot be protonated because no lone pair of electrons is available. Similarly, coordinated phosphines and arsines, like amines, cannot be protonated or quaternized. When the ligand donor atom has two or more pairs of electrons and only one of these is involved in bonding, as with oxygen of sulfur, masking of reactivity is less complete.

Other factors that become important as a result of metal bonding include changes in conformation, in oxidation-reduction potential of the ligand, and in electrical charge at the reactive sites of the ligands. These changes may alter such properties as the ease with which the ligand undergoes oxidation, protonation, or deprotonation.

EDTA is fairly effectively masked against oxidation by permanganate if the EDTA is complexed with Cr(III) or Bi(III).[17] This is also true of hydroxyethylethylenediaminetriacetic acid and its Cr(III) and Co(III) complexes towards oxidation by vanadate.[18] Oxalate coordinated to Co(III) does not undergo many of its typical reactions with oxidizing agents.[19] So, too, in the presence of a large excess of molybdic acid in sulfuric acid, oxalic acid is not oxidized by permanganate. Excess Hg(II) masks sulfite ion against oxidation by permanganate. Similarly, complexing of reducing sugars with borate masks them against oxidation by alkaline solutions of ferricyanide,[18] and glycine complexed to Cr(III) is not readily attacked by nitrite or nitrous acid.[20] In the titrimetric determination of Nb as its peroxide complex in dilute sulfuric acid, free hydrogen peroxide reacts quickly with potassium hydroxide but the Nb complex is only slowly attacked.[21]

Complex formation with Cu(II) can mask the amino or the carboxyl group of α-amino acids against further reaction. α,β-Diaminocarboxylic acids form two kinds of copper(II) complexes, depending on the pH of the solution.[22,23] In weak acid, bonding is mainly through the α-amino group and the carboxylate ion, leaving the protonated β-amino group free, whereas in neutral or alkaline solution bonding is through the two amino groups and the carboxylate ion is free. Synthetic applications of this masking, involving reactions of free terminal amino groups, as in lysine and ornithine, or of carboxyl groups, as in glutamic acid, have been discussed elsewhere,[24] together with the limitations of the method.[25]

This kind of masking is effective when terminal amino groups are to be reacted with urea, potassium cyanate, fluoronitrobenzene, and acetylating agents. A difference in rate constants of about 30-fold between the reactions of free and coordinated groups is sufficient to allow adequate masking in reactions such as these.[26]

Masking of hydroxyl reactions can occur in ligands when the oxygen atom is coordinated to the central metal ion, but in general this masking is only partial.

The rate of coupling of diazotized sulfanilic acid with the zinc chelate of 8-hydroxyquinoline-5-sulfonic acid in pH 5 acetate buffer is about one fifth of that for the free ligand.[27] So, too, coordinated thiols do not react with alkyl halides when the sulfur atoms are in bridging positions between two metal atoms in a polynuclear complex,[28] but in mononuclear complexes much of the reactivity typical of an organic sulfide is retained.[29]

In some cases reaction may proceed slowly, not because the metal complex is reacting but because the small amount of ligand existing in dynamic equilibrium in the solution is progressively removed, leading finally, to complete dissociation of the complex. This can be detected by observing whether or not the reaction rate approaches a limiting value with increasing concentration of metal ion. This problem does not arise with complexes which do not dissociate readily in solution. At pH 8 the "robust " glycerol–telluric acid complex is stable to periodate whereas glycerol is rapidly oxidized.[30] Part of the effectiveness of this masking may lie in the similar coordinating properties of tellurate and periodate ions.

The presence of a metal ion is not always necessary for ligand reactivity to be masked. Thus bisulfite reacts with iodine to give iodide ion, but this reaction does not take place if an excess of aldehyde or methyl ketone is added to a neutral solution to mask the bisulfite by forming an addition compound.

REFERENCES

1. K. B. Yatsimirskii, and T. I. Fedorova, *Dokl. Akad. Nauk. SSSR* **143,** 143 (1962); *Zh. Analit. Khim.* **18,** 1300 (1963).
2. T. I. Fedorova, and K. B. Yatsimirskii, *Zh. Analit. Khim.* **22,** 283 (1967).
3. H. A. Mottola, and H. Freiser, *Anal. Chem.* **39,** 1294 (1967).
4. H. A. Mottola, M. S. Haro, and H. Freiser, *Anal. Chem.* **40,** 1263 (1968).
5. G. Miller, and M. S. Haro, unpublished results, University of Arizona, Tucson, 1967, quoted in Ref. 4.
6. H. A. Mottola, and H. Freiser, *Anal. Chem.* **40,** 1266 (1968).
7. D. W. Margerum, and R. K. Steinhaus, *Anal. Chem.* **37,** 222 (1965).
8. R. H. Stehl, D. W. Margerum, and J. J. Latterell, *Anal. Chem.* **39,** 1346 (1967).
9. G. G. Guilbault, and R. J. McQueen, *Anal. Chim. Acta* **40,** 251 (1968).
10. M. J. Fishman, and M. W. Skougstad, *Anal. Chem.* **36,** 1643 (1964).

10a. J. S. Fritz, J. E. Abbink, and M. A. Payne, *Anal. Chem.* **33,** 1381 (1961).

10b. R. Pribil, and V. Vesely, *Chemist-Analyst* **50,** 100 (1961).

11. D. D. Perrin, *Organic Complexing Reagents,* Interscience, New York, 1964, Chap. 4.
12. F. Vydra, and R. Pribil, *Coll. Czech. Chem. Comm.* **26,** 2169 (1961).
13. F. Vydra, and R. Pribil, *Talanta* **8,** 824 (1961).
14. F. Vydra, and R. Pribil, *Talanta* **9,** 1009 (1962).
15. R. Pribil, and J. Cihalik, *Coll. Czech. Chem. Comm.* **20,** 562 (1955).
16. R. Pribil, V. Simon, and J. Dolezal, *Chem. Listy* **46,** 88 (1952).
17. M. T. Beck, and O. Kling, *Acta Chem. Scand.* **15,** 453 (1961).

18. M. M. Jones, D. O. Johnston, and C. J. Barnett, *J. Inorg. Nuclear Chem.* **28**, 1927 (1966).
19. M. M. Jones, and W. A. Connor, *Ind. Eng. Chem.* **55,** 15 (1963).
20. R. W. Green, and K. P. Ang, *J. Am. Chem. Soc.* **77,** 5483 (1955).
21. A. K. Babko, I. G. Lukanets, and B. I. Nabivanets, *Zh. Analit. Khim.* **20,** 72 (1965).
22. A. Albert, *Biochem. J.* **50,** 690 (1952).
23. R. W. Hay, P. J. Morris, and D. D. Perrin, *Austral. J. Chem.* **21,** 1073 (1968).
24. J. F. W. McOmie, *Adv. Org. Chem.* **3,** 191 (1963).
25. A. C. Kurtz, *J. Biol Chem.* **180,** 1265 (1949).
26. M. M. Jones, *Adv. Chem. Series (Am. Chem. Soc.)* **62,** 229 (1967).
27. K. E. Maguire, and M. M. Jones, *J. Am. Chem. Soc.* **85,** 154 (1963).
28. D. H. Busch, J. A. Burke, D. C. Jicha, M. C. Thompson, and M. L. Morris, *Adv. Chem Series (Am. Chem. Soc.)* **37,** 125 (1963).
29. R. V. G. Ewens, and C. S. Gibson, *J. Chem. Soc.* **1949,** 431.
30. M. M. Jones, and W. L. Lockhart, *J. Inorg. Nucl. Chem.* **28,** 2619 (1966).

CHAPTER

12

INDUSTRIAL APPLICATIONS

It is difficult to exaggerate the continuing importance to a wide range of industrial and manufacturing processes of the ability to mask metal ions so as to prevent the formation of turbidities or precipitates, the development of rancidity, off-flavors and odors, the decomposition of materials, or the appearance of unwanted or defective colors. No major developments have taken place in this field since the reviews by Chaberek and Martell,[1] and by Smith,[2] which cover the period prior to 1956, and either of these still serves as a useful introduction. Accordingly, the present Chapter has concentrated mainly on reviewing the literature since 1956 and on classifying within a general framework the examples that are quoted.

Metal ions occur adventitiously in much of the raw material used in industry, where they often produce undesirable effects in manufactured goods. The commonest of these effects are the following:

1. The formation of slightly soluble salts or hydroxides, leading to hazes, turbidities, and deposits;
2. The catalysis of chemical reactions, such as those that produce aerial oxidation of rubber, rancidity in fats, and decomposition of hydrogen peroxide;
3. The formation of unwanted colored species, causing streakiness or variation in color in different batches of commercial dyes, and the development of discolorations in a wide range of consumer products.

One or more of these effects, and others such as undesirable odor formation, can occur in the same types of materials. Successful masking procedures suppress all of these troubles, usually leading to a more attractive, more stable preparation, often with increased activity.

In most applications of masking for industrial purposes, the chemical problems involved are not very great because selective masking is not usually required. When it is, there is frequently an adequate difference in complex-forming ability between the cations to be masked and those for which activity is to be retained. Apart from the pH of the solution and the quantities of ions to be masked, the main factors that have to be considered are usually those of cost, stability (to air, heat, pH), nature of the interfering metal ions and, in some cases, biodegradability or suitability of the masking agent for human consumption. Only a limited range of masking agents currently finds much use in industry, and these comprise polyphosphates, aminopolycarboxylic acids such as EDTA,[1] and salts of organic hydroxy and polybasic acids such as citric and gluconic acids. The advantages and disadvantages of each of these types can be briefly summarized.

POLYPHOSPHATES

The earliest sequestering agents used for water softening were sodium polyphosphates in which the phosphate units are linked in chains and

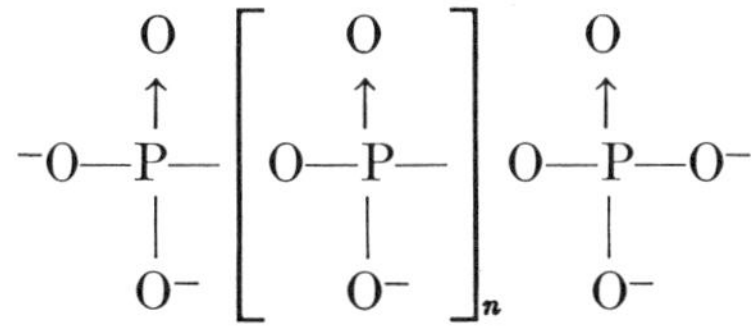

sodium metaphosphates where the ends of the chains are joined to form cyclic structures. The simplest members of the open chain phosphates are sodium pyrophosphate ($n = 0$) and sodium tripolyphosphate ($n = 1$). Much longer chains are present in the sodium phosphate glasses, introduced as water softeners in 1932,[3] but all of the polyphosphates can be used in this way to form soluble complexes with alkaline earth metal ions.[4]

The polyphosphates are still among the cheapest and most widely used sequestering agents. They are applied in near-neutral solutions to mask calcium and magnesium ions: a major example is their use to avoid the formation of insoluble alkaline earth soaps with detergents. Their effectiveness improves markedly as the temperature is raised.[5] Sequestration of iron(III) by polyphosphates at room temperature is best at about pH 4,[6,7] decreasing as the pH increases from 4 to 11 or as the chain length of the polyphosphate is increased.[7]

Published values[8] of stability constants for metal complexes with polyphosphates vary widely and no reliable conclusions can yet be drawn regarding the effect of increasing chain length on the strengths of the resulting complexes. However, results for the magnesium and calcium complexes of adenosine mono-, di-, and triphosphate (for Mg, log $K_1 = 1.7, 3.1, 4.0$, respectively; for Ca, 1.4, 2.8, 3.6 at 20° and $I = 0.1$)[9]

[1] In the following discussion, unless otherwise indicated, the abbreviation EDTA may stand for either the free acid or its sodium salts. Because of their much better solubility in water, EDTA is ordinarily added as one of its sodium salts.

suggest that tripolyphosphate would form stronger complexes than pyrophosphate which, in turn, would be a better complexing agent than orthophosphate. This can be rationalized in terms of increased polydentate ring formation with increasing numbers of phosphate groups. Much smaller gains in stability would be expected with further increase in chain length. Measurements of the sequestering action of the simpler polyphosphates[10] are consistent with this interpretation. With sodium polyphosphates and calcium ion, effectiveness increases with chain length up to 5 to 10 phosphate groups but then does not change further.[11] Shorter chain lengths are better for sequestering magnesium and iron(III).[11]

Technical aspects of the manufacture and use of polyphosphates have been discussed elsewhere.[12]

Unfortunately, polyphosphates tend to hydrolyze to the orthophosphate anion, which forms very insoluble salts with alkaline earth metal ions. The rate of this "reversion" depends on pH, temperature, and chain length of the phosphate. It is also catalyzed by metal ions, and the catalytic effect probably depends on the electron-attracting ability of the metal ion concerned. The process involves chain degradation, hydrolysis increasing rapidly with chain length, and it is also catalyzed by hydrogen ions.[13]

These factors suggest that short rather than long chain polyphosphates are preferable as sequestering agents, especially if effects due to reversion are to be minimized.

AMINOPOLYCARBOXYLIC ACIDS

Advantages of aminopolycarboxylic acids such as EDTA, used as their sodium salts, include the very good stability of the aqueous solutions to prolonged heating. They bind alkaline earths and other metal ions much more strongly than do the polyphosphates, and their sequestering ability is constant over a wide pH range. They are also stable to alkali. Their greater cost is partly offset by the smaller amounts that are needed, especially in alkaline solution, and their ammonium and amine salts have the advantage of greater compatability with soaps and organic substances.

When strong solutions of EDTA are used, for example in the descaling of boilers, the "spent" chelating solutions can be regenerated by adjusting to pH 3 with sulfuric acid to precipitate calcium sulfate which is filtered off, and the pH of the solution is again brought to pH 6–8.[14] Alternatively, by cooling the calcium-depleted solution and acidifying to pH 1–1.5, the EDTA can be precipitated as the free acid and collected.[14,15]

EDTA continues to be the most widely used member of this group, but a steadily increasing number of derivatives showing some variations in metal-binding ability is now available. These include 1,2-diaminocyclohexane-*N*,*N*,*N'N'*-tetraacetic acid (CDTA), diethylenetriaminepentaacetic acid (DTPA), ethylene glycol-bis(β-aminoethyl ether)-*N*,*N*,*N'*,*N'*-tetraacetic acid (EGTA), and *N*-hydroxyethylethylenediaminetriacetic acid (HEDTA). In most cases their greater cost is a disadvantage. Nitrilotriacetic acid is a simpler, but related, sequestering agent. It is a suitable alternative to polyphosphates for use as a chelating agent in detergents.[16]

EDTA has poor masking ability for iron at high pH values and in caustic soda solutions, whereas triethanolamine is effective under these conditions. A sequestering agent combining both sets of properties has been proposed in which 1–3 of the amino hydrogens in ethylenediamine are replaced by hydroxyethyl groups and the remainder by carboxyl groups.[17]

HYDROXY ACIDS

In aqueous solutions the anions of α-hydroxy acids are useful sequestering agents because they form stable chelate complexes with metal ions. These complexes are of two kinds. If the solutions are neutral or weakly alkaline, bonding of the metal ion is by electrostatic interaction with carboxylate ions and by coordination to α- (and possibly β-) hydroxyl groups. With citrate ion this involves two carboxylate ions and one hydroxyl group, leading to the formation of a terdentate complex containing a 5- and a 6-membered ring. Similarly, gluconate ion has one carboxylate ion and one (or two) hydroxyl groups available for simultaneous chelation of a metal ion. Such considerations lead to the following order of effectiveness at pH 11 or below:

$$\text{citrate} > \text{tartrate} > \text{saccharate} \sim \text{gluconate}$$

Gluconic acid is further weakened as a chelating agent by its existence in aqueous solution in equilibrium with γ- and δ-gluconolactones.

Citrate ion finds application as a masking agent in low and middle pH ranges, and it also finds a number of related uses, such as the derusting, descaling, and cleaning of ferrous and nonferrous metals. In acid solution, oxalate ion masks aluminum, copper, and iron(III) ions, and dissolves iron oxide.

Nevertheless the stability constants of the alkaline earth metal complexes are not very high, so that appreciable excesses of the ligands must be present if masking is to be effective. Hydroxy acids are extensively used in neutral or weakly alkaline solutions when only a moderate degree of masking is required but their particular advantages as sequestering agents reside in their behavior in strongly alkaline solution.

At higher pH values protons ionize from hydroxyl groups of hydroxy acid anions to form a much more powerful chelating agent. Although at pH 11 gluconate ion is a weaker sequestering agent than pyrophosphate for calcium ion, in 3% sodium hydroxide it is more powerful than EDTA[18] or the polyphosphates, the sequence being[19] EDTA > hexametaphosphate > tetraphosphate > tripolyphosphate.

The pH region where this type of bonding becomes important depends on the pK of the hydroxy acid anion, and this, in turn, is a function of the substituents present. Prediction[20] of the acid strengths of hydroxyl groups in α-hydroxy acids leads to a sequence which is in the reverse order to the one given above, in agreement with the observation that in solutions containing more than 5% sodium hydroxide, gluconate is more effective than citrate in masking calcium against precipitation as its oxalate. It also prevents the adherence of scale in aluminum etching. In strongly alkaline solutions gluconate and saccharate ions are also good masking agents for Fe(III). Under these conditions this masking action is only slightly affected by the presence of calcium or magnesium ions. Such hydroxy acids are useful chelating agents because they are effective over a wide pH range and are nontoxic and noncorrosive. Poly-(itaconic acid) and its water-soluble salts have been proposed as sequestering agents for di- and tri valent metal ions (of Al, Zn, Pb, Cu, and Fe).[20a]

12.1 THE USE OF MASKING AGENTS TO AVOID PRECIPITATES

SOAPS AND DETERGENTS

Water from natural sources frequently contains calcium, magnesium, or other cations such as iron, manganese, or copper, which form precipitates with soap, deposit as scale in boilers, and cause haze or turbidity in otherwise clear solutions. Where large quantities of water are involved, it is usually possible to remove these cations by ion exchange or precipitation. With smaller volumes, or when the water is reused, an alternative procedure is the addition of a suitable sequestering agent to mask the chemical reactions of the metal ions without the need to separate them from the solution.

The effectiveness of such an agent is sometimes expressed in terms of its *sequestration value*, defined as the weight of sequestering agent that is needed under specific conditions, per unit quantity of multivalent positive ion, to prevent the precipitation of alkaline earth salts. However, this concept is of only limited value because it clearly depends on a number of variables such as the pH of the solution and the nature and concentrations of other species, particularly anions, that are present.

One of the best known examples in which masking agents are added to avoid precipitation is in processes in which soaps or detergents are used with hard water. The insoluble soaps formed by calcium, magnesium, and other metal ions are aesthetically unsatisfactory, represent a waste of the cleansing agent and cause interference in operations such as the bleaching and dyeing of textiles (where they cling to, or become embedded in, the fibers) or the cleaning of metal surfaces in baths containing sodium soaps and carbonate, hydroxide, phosphate, or silicate ions. For near-neutral solutions polyphosphates are effective and cheap sequestering agents, especially if the solution is not heated or stored for too long, so that alkali metal tripolyphosphates have been proposed for use in detergents to sequester calcium.[21,22] Similarly, in the presence of sodium hexametaphosphate, adsorption of iron(III) ions on to viscose rayon pulp is negligible.[23]

Foaming and detergent capacities of soaps and detergents are greatly improved by inclusion of EDTA[24,25] or nitrilotriacetic acid,[25] both of which overcome the deficiencies of soap in hard water.[26] EDTA and nitrilotriacetic acid are effective masking agents for iron in the laundering process for cotton and synthetic fibers, and for removing iron from fibers.[27] Tripoly- and pyrophosphates are also good masking agents for iron(III), especially in the presence of sodium silicate, but their binding capacity is strongly pH-dependent.[27] Incorporation of EDTA or DTPA into soaps lowers their turbidity point.[28] Sodium tripolyphosphate and EDTA, or sodium nitrilotriacetate can be used separately in liquid detergents[29,30] or in biodegradable detergents containing alkyl sulfates and sulfonates.[22]

Polyphosphates and aminopolycarboxylic acids show a well-marked synergic action, so that mixtures of sodium polyphosphates and nitrilotriacetic acid are included in some detergents and soaps.[31,32,33] They are used, for example, at pH values above 10 to prevent the precipitation of iron(III) and other metal ions during textile finishing.[34] Similarly EDTA and polyphosphates are suitable for sequestering calcium ions and for removing metal ions adsorbed on the fibers or present in the wash water when textiles and wool are washed[35] or scoured to remove oils, waxes, and grease.

Saturated solutions of detergents based on sodium metasilicate have been stabilized against precipitation by the addition of EDTA.[36] Nitriletris (lower alkylidenephosphonic acids) and their alkali metal salts have been proposed for preventing the precipitation of calcium and magnesium in alkaline cleaning solutions.[37] In strongly alkaline solutions, such as are used for bottle washing, gluconate ion is usually preferred, either alone or in admixture[38] with EDTA. An evaluation of possible sequestrants for bottle-washing in hot sodium hydroxide solution gave the following sequence:[39]

Na saccharate, glucuronate, gulonate > galactoheptonate > tripolyphosphate, lactose carboxylate > gluconate, glucoheptonate >> EDTA, and shows the usefulness of acids obtained by partial oxidation of sugars. This is the rationale behind the suggested use of oxidized molasses as a binding agent for multivalent metal ions in strongly alkaline detergent solutions,[40] and of alkali metal salts of invert sugar heptonic acids in caustic solutions used in bottle washing.[41]

Again, with the development of shampoos and other soap preparations, the inclusion of materials to mask metal ions has led to improved clarity of the products and has diminished the problem of their use with hard water.

FOODS AND BEVERAGES

Turbidities and hazes in foods and beverages, including wines and beer, are often due to the formation of insoluble salts which can be avoided by lowering the free metal ion concentration sufficiently. For example, complex formation with EDTA masks calcium that otherwise forms a calcium pectinate haze in clear jellies. EDTA can also be used to clear turbidities due to iron and other metals in wine.[42,43] Sequestering agents prepared by the reaction of polyalkylenimines with cyclic carbonates or sulfites are claimed to be suitable for use in the clarification of beer, fruit juices, and other beverages, as stabilizers in pharmaceuticals and as water softeners.[44]

The soaking of peas and beans in dilute polyphosphate solutions prior to canning or freezing makes them more tender by removing calcium and magnesium that would otherwise react with pectin present in the skins of these legumes.[45]

The thickening of condensed milk can be retarded by the addition of EDTA to combine with casein calcium.[46] Formation of crystals of struvite (magnesium ammonium phosphate) in canned sea foods can be prevented by adding sodium hexametaphosphate and EDTA in sufficient amounts, but this affects color, taste, and texture:[47] use of citric acid is more satisfactory.[48]

Tetracyclines have metal-complexing ability, so that in the presence of multivalent metal ions their antibiotic activity is diminished. Addition of a chelating agent decreased this antagonism when tetracyclines were used for shrimp preservation.[49]

The use of all sequestering agents in comestibles raises the question of possible toxicity. The United States Federal Food, Drug and Cosmetic Act sets maximum permissible limits for the concentrations of EDTA in many foods and drinks, such as vegetables,[50,51,52] sweeteners,[53] food dressings, and fats,[54] and carbonated soft drinks.[55] The poor absorption of EDTA from the gastrointestinal tract makes it unlikely that EDTA will have any serious effects within the mammalian body. On the other hand, prolonged administration of EDTA at a high level in the diet could diminish the uptake of essential divalent cations and also lead to disturbances in the alimentary canal because of interference with the activity of the microflora.

Citric acid occurs naturally in plants and animal tissues and is non-toxic at usual concentrations. Most of the other hydroxy acids used in masking can also be taken in small amounts without physiological upset, but oxalic acid and oxalates are toxic, even in dilute solutions.

Because of their highly polar character polyphosphates are not absorbed from the gut and, except insofar as they may alter the pH of digestive material or prevent the uptake of metal ions, their use does not appear to be attended with any health hazards.

INDUSTRIAL OPERATIONS

BOILER SCALE

The prolonged operation of boilers in regions where water is "hard" leads to problems in the removal of scale, resulting from the deposition of insoluble carbonates, phosphates, silicates, and sulfates, usually of iron, calcium, and magnesium. Formation of such scale can be avoided, and deposits containing these three cations can be redissolved by using EDTA[56] or some other suitable chelating agent in the feed water. When EDTA is used, it is necessary to have an oxygen scavenger present to reduce any iron(III), which hydrolyzes and is not adequately sequestered, and also to prevent any corrosion.

With this precaution, Na_4EDTA has been used in a boiler at 900 psi and 280°C, without decomposition, and with the elimination of hardness, iron-based sludge, and deposits.[57] Typically, EDTA treatment of feed water in steam boilers operating at 100 atmospheres pressure masked calcium and magnesium, and dissolved deposits on the boiler surface.[58] It also decreased the iron content of the steam.[59]

The performance characteristics of EDTA and nitrilotriacetic acid in boiler treatment have been discussed[60] and many preparations containing these sequestering agents have been patented for the prevention of corrosion and boiler scale. One such class comprise aminopolycarboxylate salts, citrate ion, surfactants, and corrosion inhibitors.[61] Nitrilotriacetic acid is suitable for use up to 800 psi, whereas EDTA is effective to 1500 psi.[62] EDTA is less useful in preventing the deposition of copper in boilers unless the solution is alkaline.[63]

When sequestration of iron is important, this can be achieved by a mixture of EDTA and tripolyphosphoric acid at pH 8 and above.[64] The solution is stable at elevated temperatures. Sodium gluconate at pH 8.5 has also been proposed for preventing scale formation in steam boilers, evaporators, and stills.[65] When boiler scale is due mainly to carbonates it can be removed by malic, malonic, tartaric, citric, gluconic, lactic, or salicylic acid in the presence of an organic sludge conditioner.[66]

OIL WELLS

The use of hydroxy and polycarboxy acids for masking iron is well established. This is the basis of the suggested use of acidified citric acid/acetic acid mixtures as sequestering agents for oil wells containing deposits of iron salts and oxides.[67] Mixtures of glycolic, diglycolic, citric, and tartaric acids are also claimed to prevent scale formation in oil wells.[68]

DERUSTING AND DEGREASING

Similarly, the addition of sodium gluconate to alkaline baths provides a nontoxic replacement for sodium cyanide in the alkaline electrolytic degreasing of metals at room temperature and is also suitable for use in hot alkaline solutions for derusting steel[69,70] and degreasing iron and other metals.[69,70,71,72] Ferric hydroxide redissolves in 3% sodium hydroxide solutions of sodium gluconate.[73] The use of alkaline sodium gluconate solutions containing a reducing agent has been suggested for the removal of heavy metal soaps, such as zinc stearate, that are used to lubricate aluminum.[74] In a comparison of triethanolamine and sodium gluconate for rust removal, the former was better in 6% sodium hydroxide, the latter in 25% sodium hydroxide, and their efficiency was improved if the temperature was raised.[75] Glucoheptonate can replace gluconate for sequestering iron and removing rust.[76]

Here, again, alternative masking agents are available for the removal of iron oxide from steel. One procedure uses sodium or potassium tripolyphosphate, with a water-soluble dithionite as a reducing agent.[77] Another detergent contains EDTA with hydroxyethyliminodiacetic acid, oxalic, and citric acids, together with triethanolamine if calcium, magnesium, and copper oxides have also to be removed.[78] A recent mixture for cleaning and pickling steel includes EDTA with one or more of the following acids as free acids, salts or esters, citric, tartaric, lactic, hydroxyacetic, gluconic, heptagluconic, or methyltartronic.[79]

INHIBITION OF CORROSION

Corrosion of aluminum is inhibited in weakly alkaline solutions by the presence of alizarin derivatives which form insoluble chelates on the surface of the aluminum: The corresponding sulfonic acid derivatives cannot be used because the chelates are soluble.[80] Conversely, in the presence of fluoride ion citric acid corrodes aluminum in neutral or acidic media because the oxide film is destroyed by fluoride ion and the resulting aluminum fluoride film is soluble in the citrate solution.[81]

A chromium(VI) compound and a sequestering agent such as EDTA, pyrocatechol, or sorbitol inhibit the corrosion of ferrous metals in contact with aqueous solutions of alkalis.[82] A similar mechanism probably explains the use of EDTA or nitrilotriacetic acid to inhibit the corrosion of stainless steel trays by bleaching solutions, containing ferricyanide, that are used in photographic processing.[83]

MILKING MACHINES

Sodium hexametaphosphate is claimed to be the best sequestrant for washing milking machines not made of stainless steel or glass.[84] Use of EDTA alone[85] or in admixture with hexametaphosphate[84] greatly increased the corrosion loss of aluminum and tinned copper. Nevertheless EDTA is used in cleaning milking equipment to avoid deposition of calcium soap or inorganic calcium salts,[86,87] where its main advantage over sodium hexametaphosphate lies in an improved performance with hot solutions.[87] Use of EDTA in bottle-washing detergents rapidly removes calcium phosphate deposits from milk bottles.[87]

ELECTROPLATING

The use of masking agents is also important in electroplating, especially with alkaline plating baths, where they not only retain impurities in solution but they also prevent the precipitation of the relatively high concentrations of the desired metal ions. The complexing agents present in such systems also fulfill other functions related to the efficiency of the plating process and the quality of the finished product. Immersion plating of metals such as cadmium, copper, and nickel is qualitatively similar to electroplating, except that the complexing agent used must have a much greater affinity for the metal on which the plating is to be deposited than for the plating metal itself: for example, Cu(II)–EDTA solutions are suitable for plating copper on to steel.

The formation of a black precipitate of nickel phosphite from a nickel hypophosphite plating bath can be retarded by the addition of amino acids,[88,89] or aliphatic amines, amino alcohols, aliphatic acids, and polyalcohols.[90]

Differences in the effects of complexing agents on the oxidation-reduction potentials of aquo metal ions makes it possible to electro deposit one metal ion in the presence of another or, under suitable conditions, to apply a plating which is an alloy.

LEATHER

In the chrome tanning of leather, complexing agents are used to prevent the precipitation of hydrolyzed Cr(III) ions, but the stability constants of the resulting complexes must not be too great or the metal ion will also be masked against coordination to binding sites on the hides. Thus, at pH 4.0–4.5, EDTA is too powerful a complexing agent, whereas *N*-hydroxyethyliminodiacetic acid or *N*,*N*-dihydroxyethylglycine is satisfactory.[91]

The chromium(III) chelates of EDTA or nitrilotriacetic acid have no tanning action on pickled skin in acid solution, but are effective when applied to limed skins.[92] Aluminum, iron, zirconium, or titanium chelates behave similarly.[92]

The use of aluminum salts in the leather industry is limited by the instability of their aqueous solutions. Difficulties due to hydrolysis can be overcome by adding sodium citrate,[93,94,95] lactate,[95] gluconate,[94] glycolate,[95] or tartrate.[93,95]

PHOTOGRAPHY

Turbidity due to precipitation of calcium salts can be avoided in photographic solutions by adding pyridine-2,6-dicarboxylic acid.[96] More general masking of multivalent metal ions is possible with EDTA,[97,98] DTPA,[97] or nitrilotriacetic acid.[97] When developer solutions are made up from hard water, large amounts of sequestering agents such as EDTA, sodium hexametaphosphate, or sodium tripolyphosphate may be required, in which case some readjustment of pH may also be needed if sensitivity is not to be lost.[99] The desensitizing by bismuth of photographic silver bromide emulsions can be reversed with EDTA.[100]

MISCELLANEOUS

Many dyes, for example naphthols, form sparingly soluble salts with calcium and magnesium, unless EDTA or some other suitable masking agent is present. One such masking agent that has been proposed for preventing the deposition of metal ions in alkaline dye baths comprises a mixture of a polymeric alkali metal phosphate, triethanolamine, EDTA, and a polycarboxylic polyamino acid.[101]

Precipitates are produced by the prolonged storage of materials such as solutions of sodium hydroxide or sodium fluoride which attack the walls of the glass containers in which they are stored. These precipitates can also be prevented by EDTA. One example is the addition of aluminum chloride and EDTA to low-pH multivitamin preparations containing ascorbic acid and sodium fluoride in glass vials to prevent etching of the glass.[102] Similarly, EDTA can be used as a sequestering agent in liquid soap formulations to avoid the gradual formation of insoluble calcium soaps due to leaching of calcium from the glass containers.[103]

Applications of masking to agriculture include the addition of EDTA to ammonium phosphate fertilizer to mask iron(III) and aluminum against precipitation,[104] and the conversion of trace elements to their EDTA complexes to diminish precipitation with hard water when they are applied to soils.[105] Small amounts of metal-complexing and surface-active agents added during acidulation of phosphate rock diminish the extent to which insoluble phosphates are formed in the manufacture of superphosphate.[106]

EDTA added to mono-, di- and triethanolamine phosphates prevents the precipitation of calcium and magnesium from hard water used in automobile radiators.[107] Similarly, formation of iron(II) arsenite precipitates in alkaline arsenite solutions containing iron(II) can be avoided by adding EDTA or nitrilotriacetic acid.[108]

12.2 SUPPRESSION OF METAL CATALYZED REACTIONS

The most effective cations for the catalysis of chemical reactions are usually those of the transition metals such as iron, copper, and manganese. These may be present in raw materials used in industry, or they may be introduced by corrosion or physical wear of mechanical equipment. Other sources of metallic contamination include the breaking down of nonferrous alloys and surface coatings.

FOODS, BEVERAGES, AND PHARMACEUTICALS

Traces of heavy metal (especially transition metal) ions catalyze the aerial oxidation of fats, oils, proteins, and many other organic compounds, probably by a free radical peroxide chain reaction.[109,110,111] Thus, the autoxidation of sunflower, soybean, and other unsaturated oils has been explained by a two-step process involving the formation of R—O—O—H compounds, followed by their decomposition. The second step is metal catalyzed, so that complex-forming agents such as EDTA or tartrate ion can inhibit the reaction.[112] Often such autoxidations lead to the development of rancidity, turbidity, disagreeable taste, and odor, or other undesirable properties in such materials or in items manufactured from them, including foods and beverages. This oxidative destruction can include the loss of important food constituents such as ascorbic acid.

Masking the metal ions frequently suppresses or greatly diminishes these reactions so long as the metal chelate is not catalytically active, but the nature and permissible concentration of any masking agent have also to be considered, especially in items intended for human consumption. Because of their very low toxicity, their acceptable flavor, and the ease with which they are metabolized, hydroxy acids are used extensively as sequestering agents in foods and pharmaceuticals: EDTA also finds wide applications. Often, chelating agents are added to materials to enhance the stabilizing action of conventional antioxidants.[113] For example, an antioxidant with an oil-insoluble sequestering agent such as citric acid or ascorbic acid has been proposed for preventing the oxidation of fats and oils.[114]

Soybean oil-in-water emulsions are autoxidized more rapidly if heavy metal ions such as copper(II) are present: the rate is decreased if the metal ions are complexed by adding EDTA or other masking agents.[115] Similarly, citric, malic and tartaric acids showed an antioxidant effect for olive oil containing traces of added copper(II) oleate.[116] EDTA can be added to the salt used in hydrogenated oils, so as to mask copper and iron salts which otherwise tend to cause oxidation.[117] Rancidity of fats and oils is also retarded by using EDTA diesters.[118] The amine salts of EDTA and related compounds have also been proposed as oxidation and corrosion inhibitors in oleaginous materials such as animal and vegetable fats and oils, and also lubricating oils.[119]

Oxidized flavor in milk can be prevented by the addition of substances that chelate copper,[120] and the red color of meat is prolonged by using a phosphate buffer (to maintain pH), an agent to control oxidation potential, and a sequestering agent to complex multivalent metal ions.[121]

By chelating metal ions essential to the bacteria, polyphosphates inhibit the growth of pseudomonads on poultry meat.[122]

EDTA has been tried extensively in the preservation of fresh fish. Heavy metal ions catalyze the development of rancidity in blended fish muscle, iron(II), vanadium(II), and copper(II) being the most active.[123] By using EDTA, dephosphorylation of inosine-5′-phosphate in fish muscle is inhibited[124] and, although there is little effect on bacterial growth, the storage life of haddock is increased.[125] The main actions are to retard the development of odors due to oxidative rancidity and to reduce discoloration.[126]

Rancidity and resulting color spotting or staining do not develop in soaps if EDTA is added as a masking agent.[127,128] A synergistic mixture of 1-hydroxyethane–1,1-diphosphonic acid and EDTA has been proposed for use in superfatted toilet soap.[129] Similarly, a mixture of an antioxidant and a sequestrant has been described for improving the plasticity of soaps and cosmetics, and for prolonging the life of their perfumes.[130] DTPA and EDTA have also been recommended as chelating agents in a wide range of cosmetic creams and lotions.[131]

The stabilizing action of masking agents in most of the examples described above can be assessed by measuring the peroxide value of materials to which they have been added, after storage under appropriate conditions of temperature and humidity. In this way, for example, EDTA has been shown to be useful for masking traces of Cu(II) and Fe(III) in soybean oil.[132] EDTA is also effective in preventing the metal-catalyzed development of rancidity and putrefaction of lanolin-base products, including lotions and cosmetics, and of starch and protein sizes used in textile finishing.[133]

Thioglycolic acid solutions used in the permanent waving of hair are oxidized by air unless suitable metal-chelating agents such as the sodium, ammonium, or triethanolamine salts of gluconic acid, EDTA, nitrilotriacetic acid, CDTA, or HEDTA are added.[134] However, if iron is present, EDTA is not a strong enough chelating agent to prevent the formation of the red ferric glycolate complex. EDTA inhibits the oxidation of sodium sulfite solutions.[135] In sulfite solutions used for cellulose production, oxidation to sulfate is catalyzed by iron or copper, but the reaction is hindered by EDTA, DTPA, or N-(β-hydroxyethylethylenediamine)-*N,N′,N′*-triacetic acid.[136]

RUBBER AND POLYMERS

Similarly, the autoxidation of rubber, polymers, and plastic materials is catalyzed by metal ions, leading to loss of tensile strength, more rapid ageing, and discoloration. Use of chelating agents can prevent such effects,[137] so that EDTA can preserve natural rubber latex against oxidative degradation.[138] Rubber latex foams containing feldspar fillers have been stabilized with sodium hexametaphosphate which forms soluble complexes with calcium(II) and magnesium(II).[139] The effectiveness of mercaptobenzoimidazoles and disubstituted *p*-phenylenediamines as rubber antioxidants is enhanced by adding EDTA, disalicylidene–alkylenediamines, or 8-hydroxyquinoline, all of which sequester copper(II).[140]

Although EDTA has been found useful in preventing these reactions,[133,141] difficulty arises from the poor compatability of aminopolycarboxylic acids or their alkali metal salts with such materials. There appears to be scope for the development of EDTA derivatives in which one or more of the ethylene hydrogens is replaced by lipo-soluble groups. It is because of its solubility in organic media that salicylaldehydepropylenediimine has been used as a sequestering agent for oils and rubber.

PEROXIDE BLEACHING AGENTS

Sodium perborate, sodium persulfate, and hydrogen peroxide are commonly used as bleaching agents, either directly or added to soaps and soap powders. They depend on liberation of oxygen from hydrogen peroxide, present as such or liberated by hydrolysis of the per- salts. In alkaline solution, the base-catalyzed decomposition of hydrogen peroxide proceeds by a chain reaction mechanism in which a metallic cation acts as a catalyst.[142] The metal ions that are most active in this way include iron, copper, manganese, and other heavy metal ions of variable valence. They can be satisfactorily masked in near-neutral solution, thereby securing a slower, more uniform bleaching action, by the addition of small amounts of suitable masking agents. These include EDTA[143,144,145] CDTA,[145,146,147] nitrilotriacetic acid,[148] DTPA,[143] and sodium pyrophosphate.[145]

At higher pH values, HEDTA and *N,N*-dihydroxyethylglycine are more effective chelating agents for Fe(III) and DTPA for Cu(II). Like EDTA, these complexing agents are very satisfactory for use in this way because of the high conditional stability constants of their metal complexes. Alkylenediphosphonic acids, such as methylenediphosphonic acid, are hydrolytically stable in strong acid or alkali and hence can be used as metal-sequestering agent in bleaches that release chlorine.[149,150] They mask metal ions such as copper(II) that catalyze the decomposition of hypochlorite solutions.

During the washing and bleaching of textiles, traces of metal-containing impurities on the fabric can catalyze localized decomposition of the hydrogen peroxide. The resulting high concentrations of liberated oxygen lead to oxidation of the cellulose in these parts of the fabric, and to a lowering of the tensile strength of the fiber. An optimum, uniform rate of decomposition of sodium perborate in an aqueous bleach at 100° is achieved by using a sequestrant to control the desired level of copper(II) ion. For this purpose, nitrilotriacetic acid and methyl- and hydroxyethyliminodiacetic acids are claimed to be better than EDTA.[151] Sodium heptonate has also been suggested as a chelating agent for iron, copper, and manganese for use when fibers, especially cellulose, are subjected to Kier boiling and peroxide bleaching in aqueous alkali.[152]

Troublesome metal ions can sometimes be removed at earlier stages, such as during Kier boiling and mercerization of fabric: in these processes, carried out in solutions containing appreciable concentrations of free alkali, triethanolamine is the sequestering agent of choice for Fe(III), along with EDTA to chelate alkaline earth ions.[153] Sodium gluconate is a suitable alternative.

Similar considerations apply to the bleaching of wood pulp. EDTA,[154] DTPA,[136,155,156] and sodium tripolyphosphate (to inactivate iron compounds)[157] have been used to reduce the metal-ion catalyzed decomposition of the alkaline peroxide solutions and to obtain a whiter product. Addition of DTPA to chlorine dioxide bleaching solutions inhibits subsequent discoloration of wood pulp due to aging.[136]

MISCELLANEOUS

Many medicinals and other organic chemicals undergo oxidative attack in aqueous solution in the presence of heavy metal ions unless masking agents are present. The following are typical examples.

Ascorbic acid, in solutions and pharmaceutical preparations, is stabilized by EDTA,[158,159] together with sodium sulfite.[159] In phosphate buffers, plant polyphenols such as rutin and quercetin are antioxidants for ascorbic acid by virtue of their chelating ability for heavy metal ions.[160]

EDTA protects aqueous solutions of adenosine triphosphate and other polyphosphate derivatives of adenosine by binding traces of heavy metal ions that catalyze their dephosphorylation.[161] It also greatly reduces the deterioration of the weakly basic salt of carboxymethyl cellulose by the action of copper(II) salts.[162]

A rather unusual application is the use of Na_4EDTA and the sodium salt of *N*,*N*-diethanolglycine to diminish the copper-catalyzed oxidative decomposition of monoethanolamine solutions used as carbon dioxide scrubbers in submarines.[163]

The decomposition of flavine–adenine dinucleotide in aqueous solution is catalyzed by traces of heavy metal ions. However, their action can be masked by many agents, including EDTA, CDTA, DTPA, HEDTA, or less satisfactorily, sodium hexametaphosphate or pyrophosphate, 8-hydroxyquinoline-5–sulfonic acid, citric acid, malonic acid, malic acid, tartaric acid or nitrilotriacetic acid.[164]

Penicillin solutions are stabilized by the sodium salts of sequestering agents such as EDTA, CDTA, DTPA, or nitrilotriacetic acid.[165] EDTA suppresses the deterioration of physostigmine solutions by traces of iron.[166] Similarly, the addition of sequestering agents controls the trace metal contaminants that accelerate the decomposition of prednisoline in aqueous solution.[167]

EDTA, with or without sodium bisulfite, stabilizes sodium sulfadiazine solutions in the presence of traces of copper and iron.[168] Thioxanthene solutions are also protected by adding EDTA, DTPA, CDTA, or similar chelating agents.[169]

In the dyeing industry iron and copper catalyze the decomposition of diazotized amine in the coupling bath when dyes such as naphthol AS are used. Here again, addition of EDTA overcomes this difficulty.[133] EDTA is also effective against metal ions which catalyze the surface oxidation of sulfur dye colors.

The enzymic breakdown of guar gum, triggered by iron, is inhibited by adding iron-chelating agents such as citric, oxalic, tartaric, or phosphoric acid.[170]

By chelating copper, *N*,*N′*-bis(3-alkenyl salicylidene)-1,2-diaminopropanes stabilize gasoline against catalytic oxidation.[171]

Blocking iron with EDTA prevents the premature exhaustion of tanning baths which can arise from the iron-catalyzed condensation of customary tanning agents to form deposits.[172]

12.3 PREVENTION OF UNWANTED COLOR DEVELOPMENT

DYES AND PAINTS

Many dyes are also metal-chelating agents and the complexes they form with metals differ in color from the original dye. Metal ion impurities in dye baths can thus lead to color variation between batches or to spotting and streakiness in the dyeing of materials. This can even occur when metal-lakes are being formed if the undesirable metal ion reacts more strongly with the dye. Also traces of metal ions, present on the material or introduced from the water of the equipment, can be precipitated as insoluble soaps on the fabric during earlier processing.

These deposits can result in faults such as uneven bleaching, dyeing, and finishing. Similarly, in casein-type paints heavy metal ions can form complexes with the protein, giving rise to discoloration and also leading to coagulation of emulsions. In other cases alkaline earth or heavy metal salts of dyes or leuco dyes are insoluble, precipitating from solution and possibly adsorbing unevenly on to the fabric being dyed.

These effects can be avoided by the use of suitable masking agents which must be able to compete successfully with the dye molecule for the metal ions concerned. The high stability constants of metal–EDTA complexes and, in most cases, their lack of color make EDTA suitable for most such applications. However, in cases in which the dye itself is a metal complex the choice of sequestering agents for other metal ions is clearly limited by the requirement that the essential metal ion is not also masked. In dye baths that use acid and chrome dyestuffs pyrophosphates can be used to mask iron:[173] more recently EDTA has been recommended.[174]

OTHER INDUSTRIAL APPLICATIONS

During the manufacture of paper ferric and other metal ions can cause discoloration of unbleached sulfite pulp and paper and also reduce the strength of the latter because of localized oxidation during bleaching. EDTA, pyrophosphate, and hexametaphosphate are suitable sequestering agents for preventing these effects.

Addition of complexing agents for heavy metals during the formation of polymers and copolymers of acrylonitrile avoids yellowing.[175] Color stabilization in polyurethans is improved by adding alkylamines as chelating agents.[176] Similarly, discoloration in butadiene copolymers with styrene or acrylonitrile is diminished by including suitable chelating agents such as EDTA or nitrilotriacetic acid in the aqueous emulsion.[177] Color-free synthetic rubber can be made by adding excess EDTA to mask iron(III) present and necessary in the emulsion polymerization process.[178]

The use of alkaline detergents in the presence of manganese ions can lead to discoloration of metallic and nonmetallic surfaces. The reaction can be retarded by adding sodium gluconate.[179] Gluconic acid is also added to a water-soluble peroxymonosulfate to give a solid cleaning composition which inhibits discoloration of lead-bearing enameled surfaces.[180]

Aqueous solutions of phenolic acids and their salts can be stabilized against color formation by using deoxygenated water and small amounts of thiols such as 2,3-dimercaptopropanol, mercaptosuccinic acid, or glutathione which are both antioxidants and metal-complexing agents.[181] EDTA protects phenylephrine solutions against oxidative discoloration.[182]

FOODS AND DOMESTIC ITEMS

Trace amounts of copper and iron can lead to discoloration in vegetables unless suitable masking agents, such as EDTA, are present. Similarly, EDTA prevents canned or frozen sea foods from becoming discolored by the formation of copper, iron, and manganese sulfides.

Discoloration can also develop in soaps and essential oils as a result of metal catalysis and in sun-screening agents such as aminobenzoic acid esters, methyl anthranilate, methylsalicylate, or methyl umbelliferone unless EDTA is included in their formulations.

12.4 OTHER APPLICATIONS OF MASKING AGENTS

The decontamination of cotton cloth and other fabrics containing radioisotopes is much more effective if the laundering is done in the presence of metal-complexing agents such as nitrilotriacetic acid or EDTA,[183] or polyphosphates.[184] EDTA, sodium tri- and hexametaphosphate, oxalate, citrate, and tartrate were used for this purpose in another study.[185] Mono-, di- and triethanolamine salts of EDTA, and also a sodium EDTA salt, have been incorporated into a toilet-soap base to give a highly efficient decontaminating agent for surfaces contaminated with radioactive and heavy metal products.[186]

Incorporation of EDTA or a similar aminopolycarboxylic acid type of chelating agent in electrolytic recording paper diminishes discoloration of the paper during storage and reduces the tendency for the recording to "trail" behind the electrode.[187] Use of EDTA in the treatment of wood pulp slurry improves the resistivity of paper for the manufacture of capacitors.[188]

Partial separation of dissimilar metal ions in aqueous solution can be achieved by selectively sequestering one kind of metal ion, and then removing the other by foam fractionation in the presence of a surface-active agent. Thus, foam enrichment of strontium(II) occurs in a solution containing calcium and strontium ions if EDTA is added, followed by sodium *o*-hydroxyphenylbutylbenzenesulfonate.[189]

Addition of Na_4EDTA to oil-in-water emulsions used in lubricating aluminum and its alloys controls the free calcium(II) ion content and helps to keep the pH between 8 and 10.[190]

The astringent taste of beer treated with zinc(II) can be removed by chelating the zinc with glycine, EDTA, CDTA, DTPA, or sodium glutamate.[191] Similarly, the addition of EDTA to cider containing iron and tin removes the metallic taste and the cloudiness.[192]

A detergent with metal-sequestering properties that has been suggested for removing bloodstains comprises a mixture of an organic acid such as dodecylbenzenesulfonic acid with mono-, bis-, tris-, and tetra(hydroxypropyl) ethylenediamines.[193]

REFERENCES

1. S. Chaberek, and A. E. Martell, *Organic Sequestering Agents,* Wiley, New York, 1959, Chap. 7.
2. R. L. Smith, *The Sequestration of Metals,* Chapman and Hall, London, 1959, Chap. 7.
3. U.S. Pat. 1,956,515 (1934); Re-issue 19,719 (1935), R. E. Hall.
4. J. R. Van Wazer, and C. F. Callis, *Chem. Revs.* **58,** 1011 (1958).
5. O. Oldenroth, *Fette, Seifen, Anstrichmittel* **64,** 468 (1962).
6. T. Todani, N. Okada, and S. Maeshige, *Nippon Shokuhin Kogyo Gakkaishi* **10**(2), 62 (1963).
7. R. R. Irani, and W. W. Morgenthaler, *J. Am. Oil Chemists' Soc.* **40,** 283 (1963).
8. L. G. Sillén, and A. E. Martell, "Stability Constants of Metal-Ion Complexes," *Chem. Soc. (London) Special Publn. No. 17,* (1964).
9. A. E. Martell, and G. Schwarzenbach, *Helv. Chim. Acta* **39,** 653 (1956).
10. R. Bernstein, and H. Fleischer, *Soap Sanit. Chemicals* **26**(9), 40–41, 82, 153 (1950).
11. S. Ueda, *Kogyo Yosui* **37,** 32 (1961).
12. J. C. Harris, in *Phosphorus and its Compounds,* J. R. Van Wazer, ed., Interscience, New York, 1961, Vol. II, Chap. 28.
13. J. R. Van Wazer, E. J. Griffith, and J. F. McCullough, *J. Am. Chem. Soc.* **77,** 287 (1955).
14. M. C. Bennett, N. O. Schmidt, L. F. Wiggins, and W. S. Wise, *Intern. Sugar J.* **58,** 249 (1956).
15. U.S. Pat. 3,321,521 (1967), G. T. Kerr.
16. R. R. Pollard, *Soap Chem. Specialties* **42**(9), 58, 130 (1966).
17. Brit. Pat. 810,113 (1959), R. L. Smith, and P. Womersley.
18. S. Anastasiu, R. Iordanescu, A. Anghelescu, Z. Dumitrescu, and S. Pavel, *Rev. Chim. (Bucharest)* **17,** 607 (1966).
19. J. C. L. Resuggan, *J. Soc. Dairy Technol.* **10,** 84 (1957).
20. G. B. Barlin, and D. D. Perrin, *Quart. Revs.* **20,** 75 (1966).

20a. U.S. Pat. 3,405,060 (1968), R. P. Carter, and R. R. Irani.

21. Neth. Pat. 288,064 (1965), Monsanto Co.
22. Belg. Pat. 665,954 (1965), Proctor and Gamble Co.
23. Z. N. Naumova, and I. A. Nagrodskii, *Bumazhn. Prom.* **38**(2), 15 (1963).
24. R. Hette, *Congr. mondial detergence et prods. tensio-actifs, 1er Congr., Paris* **2,** 636 (1954).
25. R. R. Pollard, *Hydrocarbon Process.* **45**(11), 197 (1966).
26. A. K. Prince, and W. R. Merriman, *Soap Chem. Specialities* **34**(7), 39, 87 (1958).
27. O. Oldenroth, *Fette, Seifen, Anstrichmittel* **65,** 418 (1963).

28. Ger. Pat. 1,141,040 (1962), Chemische Werke Witten G.m.b.H.
29. French Pat. 1,484,489 (1967), Proctor and Gamble Co.
30. U.S. Pat. 3,265,625 (1966), E. Grob.
31. Ger. Pat. 1,202,926 (1965), Knapsack Aktiengesellschaft.
32. Belg. Pat. 670,531 (1966), Proctor and Gamble Co.
33. French Pat. 1,437,367 (1966), Proctor and Gamble Co.
34. Brit. Pat. 873,488 (1957), J. A. Benckiser G.m.b.H.
35. A. L. de Jong, *Tex.* **25,** 242 (1966).
36. French Pat. 1,328,416 (1963), G. Mounier.
37. U.S. Pat. 3,278,446 (1966), R. R. Irani.
38. U.S. Pat. 2,584,017 (1952), V. Dvorkovitz.
39. R. B. Colaric, A. J. Ballum, P. Snikeris, and J. V. Karabinos, *Soap Chem. Specialties* **33**(9), 47 (1957).
40. Ger. Pat. 1,225,329 (1966), Chemische Fabrik Hoesch K.-G.
41. U.S. Pat. 3,217,034 (1965), J. V. Karabinos, and E. J. Quinn.
42. J. Farkas, *Vinarstvi* **52,** 171 (1959).
43. J. Farkas, and E. Fiala, *Kvasny Prumysl* **2,** 157 (1956).
44. French Pat. 1,425,700 (1966), Badische Anilin- und Soda Fabrik A.-G.
45. J. W. Holmquist, C. F. Schmidt, and A. E. Guest, *Natl. Canners Assoc., Information Letter No. 1170,* Jan. 28, 1948; R. V. Murray, and G. F. Peterson, *Bull. No. 22,* Continental Can Co., New York, 1951, pp. 16–17.
46. M. Maeno, Z. Saito, F. Aradate, and S. Tanaka, *Nippon Chikusangaku Kaiho* **26,** 117 (1955).
47. E. Tanikawa, Y. Nagasawa, and T. Sugiyama, *Hokkaido Daigaku Susian Gakubu Kenyo Iho* **8,** 195 (1957).
48. Jap. Pat. 2877('61) (1961), Nippon Marine Industries Co.
49. B. A. Southcott, and J. W. Boyd, *J. Fisheries Res. Board Can.* **22,** 117 (1965).
50. *Federal Register* **29,** 14,984 (1964).
51. *Federal Register* **30,** 6,915 (1965).
52. *Federal Register* **29,** 12,364 (1964).
53. *Federal Register* **32,** 6,393 (1967).
54. *Federal Register* **29,** 2,382 (1964).
55. *Federal Register* **27,** 11,344 (1962).
56. E. J. Seyb, *Brennstoff Waerme Kraft* **12,** 9 (1960).
57. J. J. Roosen, and J. V. Levergood, *Proc. Am. Power Conf.* **27,** 790 (1965).
58. E. A. Ulrich, *Energie* **19**(6), 210 (1967).
59. T. K. Margulova, and O. I. Martynova, *Tepleonergetika* **14**(10), 23 (1967).
60. C. Jacklin, *Proc. Am. Power Conf.* **27,** 807 (1965).
61. French Pat. 1,419,662 (1965), General Dynamics Corp.
62. J. Schantz, *Hawaii. Sugar Technol. Rep.* **25,** 100 (1966).
63. R. E. Elliott, *Ind. Water Eng.* **4**(7), 26 (1967).
64. Ger. Pat. 1,244,674 (1967), J. A. Benckiser G.m.b.H.
65. U.S. Pat. 3,308,062 (1967), D. A. Gunther.
66. U.S. Pat. 3,095,862 (1963), R. A. Berner.
67. U.S. Pat. 3,142,335 (1964), W. R. Dill, and F. N. Harris.

68. U.S. Pat. 3,130,153 (1964), H. F. Keller.
69. M. J. Reidt, *Galvanotechnik* **58,** 176 (1967).
70. T. Zak, and Z. Kolanko, *Galvenotech. Oberflaechenschutz* **5,** 223 (1964).
71. M. J. Reidt, and J. L. J. Tervoort, *Metalen* **20**(5), 139 (1965).
72. Belg. Pat. 659,380 (1965), Societé Continentale Parker.
73. H. J. Ferlin, P. Snikeris, and J. V. Karabinos, *Trans. Illinois State Acad. Sci.* **51,** 79 (1958).
74. Belg. Pat. 659,648 (1965), Henkel und Cie. G.m.b.H.
75. C. Rossmann, *Galvanotechnik* **58,** 166 (1967).
76. Brit. Pat. 909,596 (1962), Pfister Chemical Works.
77. Brit. Pat. 977,112 (1964), Rohm and Haas Co.
78. U.S.S.R. Pat. 172,948 (1965), All-Union Scientific Research Institute of Chemical Reagents and Pure Chemical Substances.
79. Brit. Pat. 1,119,680 (1968), Borg Holding A.-G. and Siemens A.-G.
80. H. Mihara, and Y. Hayakawa, *Denki Kagaku* **35,** 502 (1967).
81. M. Katoh, *Corros. Sci.* **8,** 423 (1968).
82. Brit. Pat. 873,103 (1959), Diversey (U.K.) Ltd.
83. U.S. Pat. 3,256,092 (1966), P. B. Means, and V. J. Miceli.
84. W. G. Whittlestone, *Australian J. Dairy Technol.* **17,** 43 (1962).
85. W. G. Whittlestone, and P. Lutz, *Australian J. Dairy Technol.* **17,** 101 (1962).
86. C. C. Thiel, L. F. L. Clegg, P. A. Clough, C. M. Cousins, D. N. Akam, and M. Gruber, *J. Dairy Res.* **23,** 217 (1956).
87. M. Wotzilka, *Int. Dairy Congr., Proc. 17th Munich* **5,** 505 (1966).
88. U.S. Pat. 2,999,770 (1953), G. Gutzeit.
89. Brit. Pat. 785,695 (1957), General American Transport Corp.
90. U.S. Pat. 2,935,425 (1960), G. Gutzeit, P. Talmey and W. G. Lee.
91. A. J. Gard, *J. Am. Leather Chemists' Assoc.* **50,** 255 (1955).
92. Ger. Pat. 1,257,352 (1967), Badische Anilin- und Soda-Fabrik A.-G.
93. M. K. Kirakos'yants, and I. P. Strakhov, *Izv. Vysshikh Uchebn. Zavedenii, Tekhnol. Legkoi Prom.* **1964,** 96.
94. French Pat. 840,907 (1939), W. K. Nelson.
95. A. Kuentzel, and S. Rizk, *Leder* **13,** 101 (1962).
96. Belg. Pat. 551,132 (1957), Gevaert Photo-Producten N. V.
97. Neth. Pat. 6,605,717 (1966), Gevaert-Agfa N. V.
98. U.S.S.R. Pat. 110,970 (1958), I. A. Novikov.
99. G. Huebner, *Bild u. Ton* **14,** 377 (1961).
100. I. Malinovski, G. Karadzhov, and M. Todorova, *Izv. Inst. Fizikokhim., Bulgar. Akad. Nauk* **2,** 219 (1962).
101. U.S. Pat. 3,055,726 (1962), H. Rudy, and K. Schuster.
102. J. R. Mellberg, *J. Pharm. Sci.* **54,** 615 (1965).
103. H. Kranick, *Soap Sanit. Chemicals* **23,**(11), 33 (1947).
104. U.S. Pat. 3,076,701 (1963). F. C. Bersworth, and A. E. Frost.
105. Czech. Pat. 124,493 (1967), V. Lakota, and J. Bartos.
106. U.S. Pat. 3,357,813 (1967), J. Elson.
107. Ital. Pat. 632,092 (1962), Lambro S.a.r.l.

108. Ger. Pat. 1,213,824 (1966), Pintsch Bamag A.-G.
109. C. E. Frank, *Chem. Revs.* **46,** 155 (1950).
110. N. Uri, *Nature* **177,** 1177 (1956).
111. C. E. Bawn, *Nature* **178,** 775 (1956).
112. G. Vago, J. Reti, and P. Varga, *Acta Chim. Acad. Sci. Hung.* **20,** 367 (1958).
113. L. Lachman, *Indian J. Pharm.* **30**(5), 109 (1968).
114. Belg. Pat. 621,645 (1962), N. V. Chemische Fabrik "Naarden."
115. W. Flueckiger, *Gordian* **68,** 77 (1968).
116. T. Bito, and K. Aoshima, *Yukagaku* **16,** 423 (1967).
117. U.S. Pat. 3,198,608 (1965), F. G. Miller and E. A. Dittmar.
118. Brit. Pat. 1,077,458 (1967), J. R. Geigy A.-G.
119. U.S. Pat. 2,805,203 (1957), C. C. Knapp, and E. O. Forster.
120. A. Goded, *Intern. Dairy Congr. Proc., 16th, Copenhagen, Sect. A* **1962,** 528.
121. U.S. Pat. 3,154,421 (1964), M. M. Voegeli, and H. J. Gorsica.
122. R. P. Elliott, R. P. Straka, and J. A. Garibaldi, *Appl. Microbiol.* **12,** 517 (1964).
123. C. H. Castell, and D. M. Spears, *J. Fish. Res. Board Can.* **25,** 639 (1968).
124. H. S. Groninger, and J. Spinelli, *J. Agr. Food Chem.* **16,** 97 (1968).
125. H. E. Power, R. Sinclair, and K. Savagaon, *J. Fish. Res. Board Can.* **25,** 2071 (1968).
126. J. W. Boyd and B. A. Southcott, *J. Fish. Res. Board Can.* **25,** 1753 (1968).
127. Brit. Pat. 875,720 (1959), Badische Anilin-und Soda-Fabrik A.-G.
128. A. M. Ille, and O. Popescu, *Ind. Aliment.* **15,** 538 (1964).
129. Neth. Pat. 6,703,756 (1967), N. V. Unilever.
130. French Pat. 1,472,251 (1967), M. Bierre.
131. A. J. Gard *J. Soc. Cosmetic Chemists* **9,** 261 (1958).
132. A. W. Schwab, P. M. Cooney, C. D. Evans, and J. C. Cowan, *J. Am. Oil Chemists' Soc.* **30,** 177 (1953).
133. *Sequestrene,* Alrose Chemical Co., Providence, R.I.
134. Pol. Pat. 45,114 (1961), Fabryka Kosmetykow "Lechia."
135. G. V. Zavarov, *Zavodskaya Lab.* **26,** 1352 (1960).
136. A. J. Gard, *Tappi* **47,** 198 (1964).
137. L. H. Lee, C. L. Stary, and R. G. Engel, *J. Appl. Polym. Sci.* **10,** 1699, 1717 (1966).
138. Brit. Pat. 856,055 (1960), Socfin Co. Ltd.
139. U.S. Pat. 3,383,341 (1968), T. Trogdon, and H. S. Smith.
140. Brit. Pat. 785,316 (1957), B. N. Leyland, and R. L. Stafford.
141. U.S. Pat. 2,560,743; 2,560,744 (1951), C. E. Rhines.
142. K. Tabe, and T. Arata, *Shokubai* **7,** 462 (1965).
143. M. Yotsuya, and S. Keimatsu, *Kami-fa Gikyoshi* **19,** 472 (1965).
144. U.S. Pat. 3,278,445 (1966), H. A. Kerry, and A. L. Andersen.
145. U.S. Pat. 3,089,753 (1963), R. E. Meeker.
146. Brit. Pat. 866,764 (1959), "Shell" Internationale Research Maatschappij N.V.

147. U.S. Pat. 2,961,306 (1960), P. R. Johnston.
148. Brit. Pat. 863,502 (1961), U.S. Pat. 2,950,175 (1960), P. R. Johnston.
149. Neth. Pat. 6,407,365 (1965), Monsanto Co.
150. U.S. Pat. 3,297,578 (1967), M. M. Crutchfield, and R. R. Irani.
151. French Pat. 1,338,856 (1963), Cadum Palmolive.
152. Brit. Pat. 1,123,071 (1968), A. L. Barber.
153. U.S. Pat. 2,544,649 (1951), F. C. Bersworth.
154. U.S. Pat. 3,251,731 (1966), A. J. Gard.
155. R. D. Spitz, *Tappi* **44,** 731 (1961).
156. R. T. Lumb, *Pulp Paper Mag. Can.* **68**(3), T131 (1967).
157. U.S. Pat. 3,023,140 (1962), C. K. Textor.
158. C. M. Dezelic, J. Grujic-Vasic, and R. Popovic, *Tehnika* **19,** 2087a (1964).
159. D. M. Ruzhitskii, *Farmatsevt. Zh. (Kiev)* **21**(4), 40 (1966).
160. C. A. B. Clemetson, and L. Andersen, *Ann. N.Y. Acad. Sci.* **136,** 339 (1966).
161. Pol. Pat. 46,783 (1963), Tarchominskie Zaklady Farmaceutyczne Poefa.
162. Y. Ogiwara, and Y. Ogiwara, *Kami-pa Gikyoshi* **21,** 397 (1967).
163. C. H. Blackly, and H. Ravner, *NASA Accession No. N 6520039, Rept. No. AD 609888* (1964).
164. T. Komatsu, *Bitamin (Kyoto)* **23,** 1 (1961); **23,** 182 (1962).
165. Belg. Pat. 648,806 (1964), Olin Mathieson Corp.
166. V. Paraak, *Acta Chim. Acad. Sci. Hung.* **33,** 121 (1962).
167. T. O. Oesterling, and D. E. Guttman, *J. Pharm. Sci.* **53,** 1189 (1964).
168. J. L. Gupta, and H. C. Mital, *Indian J. Pharm.* **27,** 251 (1965).
169. U.S. Pat. 3,248,291 (1966), H L. Newmark.
170. U.S. Pat. 3,007,879 (1958), W. A. Jordan.
171. U.S. Pat. 3,071,451 (1963), L. Schmerling.
172. Czech. Pat. 114,306 (1965), M. Kremr and V. Hrabovsky.
173. P. Rabe, *Melliand Textilber.* **28,** 352 (1947).
174. *Technical Data Sheet, No. 707012,* Cowles Chemical Company, Cleveland, Ohio.
175. Ger. Pat. 1,152,262 (1963), Farbenfabriken Bayer A.-G.
176. French Pat. 1,441,466 (1966), B. F. Goodrich Co.
177. Ger. Pat. 1,256,415 (1967), VEB Chemische Werke Buna.
178. Ger. Pat. 1,130,597 (1962), Chemische Werke Huels A.-G.
179. U.S. Pat. 3,303,104 (1967), F. K. Rubin.
180. Belg. Pat. 633,959 (1963), E. I. du Pont de Nemours et Cie.
181. French Pat. 1,438,158 (1966), E. L. Leroi, and J. Renault.
182. G. B. West, and T. D. Whittet, *J. Pharm. and Pharmacol.* **12** (Suppl), 113T (1960).
183. D. G. Stevenson, *Proc. Congr. Intern. Federation Soc. Cosmet. Chemists, 2nd, London* **1962,** 241.
184. F. Reiff, K. Schuster, and H. Heinen, *Atompraxis* **9,** 58 (1963).
185. H. Hotta, Y. Wadachi, and H. Fukata, *Nippon Kagaku Zasshi* **80,** 37 (1959).
186. U.S. Pat. 3,026,265 (1962), G. A. Campbell, and D. K. Howard.
187. Brit. Pat. 1,068,874 (1967), M. Alden.

188. U.S. Pat. 3,143,458 (1964), C. A. Charron.
189. U.S. Pat. 3,054,746 (1962), E. L. Gaden, and R. W. Schnepf.
190. Brit. Pat. 1,118,224 (1968), Dow Chemical Co.
191. U.S. Pat. 3,202,515 (1965), I. Stone.
192. B. Revis, S. M. Jafar, and W. B. Date, *Res. Ind.* (*New Delhi*) **12,** 9 (1967).
193. U.S. Pat. 3,398,097 (1968), P. W. Kersnar, and S. Taormina.

CHAPTER

13

BIOLOGICAL APPLICATIONS

Metal ions play many important roles in biological processes, particularly in maintaining osmotic balance in multiphase systems and as necessary components in a wide range of enzymic reactions. Essential metal ions in the mammalian body include those of sodium, potassium, magnesium, calcium, and traces of iron, cobalt, copper, zinc, manganese, and molybdenum. Animal and plant tissues contain many different kinds of complex-forming species, such as amino acids, peptides, proteins, carboxylic acids, and phosphates, so that each of the tissues of a living system can be likened to an arena in which these complexing species compete for the various kinds of metal ions that are present. Each of these arenas is surrounded by a barrier which can be penetrated with greater or less ease, depending on the particular ion or molecule. A neutral molecule such as 2,3-dimercaptopropanol passes more easily through a layer of fatlike molecules than, for example, a highly ionized species such as an EDTA anion.

Currently, work is in progress on multiple equilibria in assemblages of metal ions and complexing species, considered as models for biological systems.[1,2] Possibilities inherent in this approach, which uses electronic digital computers, include approximately quantitative estimates of the following:

1. Free metal ion concentrations;
2. Concentrations of all complexes, hence information as to how each metal ion distributes itself among the complexing species;
3. The effects of pH changes on metal ion distribution;
4. The effects of metal ion deficiencies or excesses (including poisonings) on the distribution of metal ions among complexing species;
5. The effects of variations in concentrations of complex-forming species;
6. The effects of adding a complexing agent; for example, a therapeutic substance such as penicillamine or 2,3-dimercaptopropanol which is intended to remove or lower the concentrations of undesirable cations, usually with a view to facilitating their excretion from the organism.

In principle this approach can ultimately be extended, by taking into account relative volumes of different tissues and distribution coefficients for species across tissue boundaries, to approximate to conditions in intact animals and humans. Consideration must also be given to the ease with which species can be modified or destroyed by enzyme action. In this way a quantitative approach appears to be possible to the design of masking agents for use in cases of poisoning by metals including lead, copper, mercury, or cadmium. Until now the choice of complexing agents for such use has been largely empirical.

There is a fundamental difference between poisoning by radionuclides and poisoning by other types of metal ions. With the latter, masking to prevent retention in the body, is the important step, and excretion is of much less urgency: in fact, chemical inactivation and irreversible deposition in deep tissue or bone may be quite satisfactory. With a radionuclide, on the other hand, radiation damage continues to be done so long as the material is present, so that the major problem becomes one of rapid and, if possible, complete removal from the body.

13.1 MEDICAL APPLICATIONS OF MASKING

The use of chelating agents to sequester metal ions provides a rational approach to the treatment of diseases resulting from metal ion excesses or deficiencies, and a large literature has grown up about such medical applications.[3] The major problem of chelation therapy lies in the need to mask or remove toxic metals such as arsenic, lead, or mercury, or to lower to normal levels metal ions such as Fe^{3+} or Cu^{2+} when these are present in physiologically abnormal amounts without, at the same time, diminishing the concentration of other essential metal ions below safe limits. Injection of the sodium salt of EDTA to mask lead can lower the level of blood calcium sufficiently to lead to tetany: prolonged administration of small amounts depletes the calcium content of the bones.

Metal poisoning may be acute or chronic and this, in turn, is reflected in the distribution of toxic cations within the body. In acute poisoning when the poison has been taken by mouth, usual locations of excessive concentrations are the gastrointestinal tract, the circulation and the soft tissues. Other means of entry are via the lungs or through the skin. Chronic poisoning frequently leads to deposition in particular tissues, such as lead in bone, or excessive copper in liver, brain and kidney. Treatment with a complexing agent will have different effects, depending on these distributions and the ease with which the metal ion can be translocated from one tissue to another, either as the metal or as its complex. This translocation may be undesirable. For example, treatment of lead poisoning with 2,3-dimercaptopropanol may result in deposition of lead in the brain tissue, and to an increased severity of encephalopathy.[4,5]

Typically, the initial result of treatment with a chelating agent will be the excretion, as a metal chelate, of much of the toxic metal present in the bloodstream: subsequent removal from other parts of the body will depend on the rate at which the metal redistributes itself within the body tissues. In general, the sooner chelation therapy is commenced after poisoning, the more effective it is.

Again, there are many ways by which the chelating agent can be administered, including oral, subcutaneous, intraperitoneal, inhalation, and intravenous routes. The particular technique chosen depends mainly on the speed with which the chelating agent must react with the toxic metal ion and the ease with which the agent and its metal chelate can pass across tissue boundaries. Because most chelating agents are rather unselective in their reactions with metal ions, it is frequently necessary to administer preparations containing essential metal ions after therapy involving metal binding. Thus, therapy with calcium disodium EDTA leads to marked depletion of essential zinc reserves in the body.[6]

The four commonest agents for the masking and removal of metal ions are the following:

EDTA, and its alkali and alkaline earth metal salts
2,3-Dimercaptopropanol and related thiols
Penicillamine
Desferrioxamine-B

Their main applications lie, respectively, in the treatment of chronic lead poisoning, arsenic and mercury poisoning, Wilson's disease, and acute poisoning with ferrous salts. It is of interest to note that these ligands represent four different types of chemical bonding. In EDTA complexes the metal is held by oxygen and nitrogen. In dimercaptopropanol two sulfur atoms are involved. In penicillamine bonding is to sulfur, nitrogen, and oxygen, whereas in desferrioxamine only oxygen-containing groups take part.

EDTA

EDTA, as the free acid or as the mono-, di-, tri-, or tetrasodium salt, complexes strongly with calcium ion, so that these forms are unsuitable for therapeutic purposes except when depletion of calcium levels is sought. This difficulty is avoided by using calcium disodium EDTA, which does not affect body calcium levels but readily exchanges its own calcium for lead or other heavy metal ions to give stable, water-soluble complexes of the heavy metals, thereby facilitating their excretion. Since its first use in this way for the treatment of lead encephalopathy[7] $CaNa_2EDTA$ has become the standard method for the treatment of lead poisoning.

In acute cases, intravenous injection of $CaNa_2EDTA$ leads to a rapid excretion of the lead complex into the urine. This removes much of the lead present in the blood and other readily accessible stores. Once the acute phase has passed, weekly injections over a prolonged period have been recommended, to allow time for redistribution of lead from less accessible stores within the body and to keep the extent of disturbance of other mineral metabolism to a minimum.[8] $CaNa_2EDTA$ should not be administered orally in such treatments, partly because it is only poorly absorbed by this route and partly because there is a danger that lead already excreted into the gut will be reabsorbed as its EDTA complex, increasing the lead burden in the body.[8] However, intramuscular or intraperitoneal injections are satisfactory.

EDTA is not readily metabolized, so that most of the injected $CaNa_2EDTA$ is finally removed in the urine, nor is it very toxic, and a daily intravenous injection of 3 g/day in man is usually well tolerated.

The amount of lead eliminated in the urine after intravenous injection of $CaNa_2EDTA$ has been found to be proportional to the lead present in the soft organs,[9] and the mechanism for its mobilization has been discussed.[10] Case histories in which $CaNa_2EDTA$ was used intravenously to treat lead poisoning have recently been given,[11] and similar treatment has been applied to calves.[12]

The nonselectivity of complexing agents in such treatments is shown by the increased excretion of copper and manganese when lead- and mercury-poisoned patients were treated with $CaNa_2EDTA$ and unithiol.[13]

In an attempt to improve the effectiveness with which lead is removed other ligands related to EDTA have been tried.[14] Of these, the most promising is DTPA (diethylenetriamine pentaacetic acid) but claims are conflicting. It has been reported as better than,[15] equal to,[16] and slightly less satisfactory than[17] EDTA for this purpose.

Advantage has been taken of the poor absorption of EDTA from the alimentary tract to prevent bloat in ruminants by reducing the concentrations of free magnesium and calcium ions in the rumen.[18] Similarly, growth depression in turkeys, caused by a high uptake of manganese, zinc, and copper, was overcome by adding EDTA to the diet to mask these metals.[19] Toxicity of vanadium in a diet fed to chickens was prevented in the same way.[20] Orally administered EDTA and DTPA also reduced radioiron absorption in man.[21] DTPA has been used similarly in treating Wilson's disease to diminish the uptake of copper from the gut.[22]

The successful use of $CaNa_2EDTA$ in the treatment of lead poisoning has encouraged its trial as a masking against other undesirable metal ions. Subcutaneous injection of $CaNa_2EDTA$ was able to antagonize the toxic action of cobalt taken orally.[23] Calcium salts of EDTA and DTPA have been suggested for intraperitoneal injection to treat cadmium poisoning, but the time between poisoning and initiation of treatment must be less than one day because the therapy depends mainly on intercepting the cadmium while it is still in the plasma.[24] When the cadmium was administered subcutaneously to mice the effectiveness of calcium sodium salts lay in the sequence[25]

> DTPA > TTHA (triethylenetetraamine hexaacetate) > EDTA > hydroxyethylethylenediamine triacetate > CDTA (cyclohexanediamine tetraacetate).

Injection of the zinc complexes of EDTA and CDTA into cadmium-hypertensive rats removed the cadmium and lowered the blood pressure.[26] Likewise, $ZnNa_2EDTA$ and $MnNa_2EDTA$, injected into rats, promoted the excretion of radio-isotopes of zinc and manganese, respectively.[27] Injection of $CaNa_2EDTA$ or $CaNa_3DTPA$ into rats caused a transient depletion of zinc, taken mainly from the liver. $CaNa_3DTPA$ was much the more effective.[28] TTHA is better than DTPA for eliminating radionuclides of cerium, yttrium, thorium, and plutonium, but they cause damage to kidney tissue.[29] The use of EDTA and DTPA to eliminate these, and similar, radionuclides has been reviewed.[30]

Nevertheless, the extent of such possible applications is limited. Thus, although DTPA is reported to be more effective than EDTA for the purpose,[31] radioactive strontium cannot be removed satisfactorily from bone with EDTA-type ligands because there is insufficient difference between the stability constants of strontium and calcium complexes: a factor of at least 50 times greater for the strontium complex is necessary. This is not, in itself, a sufficient guide in choosing masking agents for other cations in systems as diverse as biological ones; for example, EDTA forms much more stable complexes with mercury than with zinc, whereas administration of EDTA can be used to eliminate zinc but not mercury from the body. Similarly, as pH 5–9 the uptake of zinc by bone is greatly reduced if EDTA is present.[32]

EDTA has been used as a sodium salt to chelate calcium in the treatment of lime burns of the cornea,[33] and to diminish calcium concentrations in hyperglycemic patients.[34] It may also be useful as a means of altering the ionic calcium, potassium ratio in conditions such as digitalis-induced cardiac arrhythmias:[35] citrates can be used in the same way.[36] The same principle is inherent in the suggestion that Mg–EDTA might help in the relief of hypertension, by lowering the concentration of calcium ion in the blood stream as a result of exchange with the magnesium present in the Mg–EDTA complex.[37] Therapy for acute ferrous sulfate poisoning has included the intravenous injection of $CaNa_2EDTA$,[38] although, subsequently, DTPA was reported to be better than EDTA for binding iron.[39] EDTA has been incorporated in pastes and creams for protection of skin in contact with heavy metal salts[40] and for decontamination of skin which has been contaminated by radioactive cations.[41]

DIMERCAPTOPROPANOL AND RELATED THIOLS

2,3-Dimercaptopropanol (Dimercaprol, British Anti-Lewisite, BAL) was developed during World War II for use as an antidote to the arsenical poison gas Lewisite,[42] its action being based on the formation of a five-member chelate ring through the sulfhydryl sulfurs and the arsenic atom. Subsequently, dimercaptopropanol has also proved effective against acute mercury poisoning, as well as poisoning by arsenic, antimony, or gold.[43] The synthesis, chemical, and biochemical properties, and early clinical applications of dimercaptopropanol have been reviewed elsewhere.[42,44]

Dimercaptopropanol and similar dithiols act by providing additional -SH groups which compete for heavy metal ions that would otherwise combine with essential -SH groups of enzymes such as succinoxidase and the pyruvic oxidase of brain. The dithiols used in this way are all able to form very stable chelates with the heavy metal ions, liberating essential -SH groups, such as those of lipoic acid which is a cofactor in the above enzyme systems. The main disadvantages associated with the use of dimercaptopropanol are that it smells very unpleasant, is easily oxidized and is unstable in aqueous solution. On the other hand, dimercaptopropanol has poor complexing ability for alkaline earth cations and, unless it is chelated, it is rapidly degraded in the body, so that depletion of essential metal ions is not observed. The material is available as a solution in oil (which leads to slower absorption following injection, a more prolonged action, and less danger of exceeding toxic dose levels). For these reasons dimercaptopropanol must be injected intramuscularly or, in suitable cases, applied topically. Undesirable side effects, such as nausea, vomiting, and pain, are fairly common.

Nevertheless, dimercaptopropanol appears to remain the best masking agent for arsenic, and there has been no major change in its method of use, as described, for example, in the treatment of arsenic poisoning in animals[45] and humans.[46,47] Applied as an ointment, dimercaptopropanol is absorbed through the skin and is suitable for the management of dermatitis due to arsenic and gold.[48,49]

Dimercaptopropanol has been used for treating acute mercury poisoning,[47,50] although in chronic poisoning with organic mercurial compounds it may result in an increased mercury concentration in brain, liver, and muscle.[51,52] The effect of dimercaptopropanol on urinary excretion and brain mercury content has been studied,[53] and it has been used for the renal elimination of radioactive mercurials in man.[54] Dimercaptopropanol gives a more rapid initial excretion but $CaNa_2EDTA$ (intraperitoneally) has a more sustained effect.[55] Although dimercaptopropanol as ordinarily administered masks inorganic and organic mercury compounds,[56] it is ineffective orally.[57] *N*-acetyl-D,L-penicillamine and $CaNa_2EDTA$ given by mouth to pigs also failed to increase the excretion of mercury.[57]

Soon after it became available dimercaptopropanol was reported to be an effective antidote in acute poisoning of rabbits by antimony, bismuth, chromium, nickel, and mercury, but not thallium, selenium, or lead.[58] Later, thallium poisoning was also treated successfully, but fairly high levels of dimercaptopropanol were necessary.[59,60] Because of its toxicity, dimercaptopropanol does not appear to be very suitable for this purpose. $CaNa_2EDTA$ has become the drug of choice for treating lead poisoning, but the use of a mixture of dimercaptopropanol (intramuscularly) and $CaNa_2EDTA$ has been claimed[61,62,63] to be better in treating lead encephalopathy in children because these substances differ in their relative rates of diffusion to body tissues. The use of chelating agents in the treatment of lead poisoning in children has recently been reviewed.[62]

Dimercaptopropanol removes cadmium from the body, but it cannot be used for therapy because its cadmium complex causes damage to the kidney.[64] It has been applied to the treatment of cases of acute copper poisoning,[65] and of Wilson's disease,[66] the rate of removal of copper from ceruloplasmin by chelating agents being in the sequence[67]

$$\text{BAL} > \text{EDTA} > \text{diethyldithiocarbamate} >> \text{penicillamine.}$$

Many other potentially useful dithiols have been synthesized, but most of these are too toxic for clinical use. One of the most promising is disodium dimercaptosuccinate which is claimed to be less toxic to mice than dimercaptopropanol or unithiol when it is injected as an antidote in arsenic and mercury poisoning.[68] It increased lead excretion and diminished the lead content of the liver[69,70] and was comparable with $CaNa_2EDTA$ in promoting the excretion of lead from lead poisoned rabbits.[69] It has been suggested as a detoxicant for cadmium as well as lead[71] because it increased the excretion of cadmium in the urine, and reduced the mortality rate, in rabbits poisoned with cadmium.[72]

Dihydrothioctic acid is almost as effective as dimercaptopropanol as an antidote in arsenic poisoning, but it is freer of side effects and easier to administer.[73] Unithiol has also been used in the treatment of arsenic and mercury intoxication.[74] Potassium methyl- and ethylxanthates are good antidotes in poisoning with mercury(II) ion, and are claimed to be less toxic than dimercaptopropanol.[75] Sodium diethyldithiocarbamate has been suggested for the same purpose.[76] The general usefulness of thiols for removing mercury from the body is shown by the fact that dimercaptopropanol, cysteine, β-mercaptoethylamine and thioctic acid (and also EDTA) increased the renal elimination of mercurial diuretics in rabbits.[77] β-Mercaptoethylamine has also been suggested for treating thallium poisoning, because of the advantages of lower toxicity and of the ability to be administered intravenously.[78]

PENICILLAMINE

L-Penicillamine, β,β-dimethyl cysteine, was detected in the urine of patients with severe liver damage who were receiving parenteral penicillin.[79] The enhanced stability of penicillamine, compared with cysteine, prompted its trial as a chelating agent to promote the excretion of copper in patients with Wilson's disease (hepatolenticular degeneration in which intracellular deposition of copper in liver and brain is a characteristic feature).[80] Subsequently, D-penicillamine proved to be less toxic,[81,82] and it has since come into widespread use for this purpose. Penicillamine has the advantages over dimercaptopropanol that it is water soluble and stable, can be given by mouth, is tolerated in larger doses and is rapidly excreted. Currently there appears to be no satisfactory alternative treatment for Wilson's disease, and many patients have received D-penicillamine daily for months or years.[83]

Recently it has been suggested that medication with penicillamine can prevent the onset of Wilson's disease in asymptomatic subjects who, for genetic reasons, would be expected to be susceptible.[84] Justification for such a prophylactic measure has been disputed.[85,86,87]

Penicillamine has been used in the treatment of inorganic lead poisoning.[88,89] The lead is kept metabolically inactive by the penicillamine even although further doses may be necessary for its excretion. Oral penicillamine is slightly less effective in securing the elimination of lead than penicillamine given intravenously and this, in turn, is slightly less effective than $CaNa_2EDTA$ given intravenously.[89]

N-Acetyl-D,L-penicillamine has been suggested for the treatment of mercury poisoning.[90]

DESFERRIOXAMINE B

Iron-chelating substances occur naturally in certain microorganisms, and an iron chelate, ferrioxamine B, was isolated in 1960 from *Streptomyces pilosus*,[91] and subsequently synthesized.[92] Other ferrioxamines, or sideramines, are also known, occurring only in bacteria, where it is thought that they play a fundamental role in the enzymic incorporation of iron into the porphyrin skeleton. All of the sideramines have three hydroxamic acid groups as iron-binding centres, so that the central iron atom is held in a six-coordinate structure. Their stability constants are very high,[93] and desferrioxamines derived from them when the iron is removed by acid treatment have a powerful chelating ability for iron(III) but only a relatively weak affinity for most other cations.

Most of the iron in plasma is carried on the protein, transferrin, which serves to keep the free iron content in the blood below toxic levels while, at the same time, making the iron available for erythropoiesis by immature red blood cells. Under physiological conditions, desferrioxamine can attack depot iron, and to some extent transport iron, but not porphyrin iron.[94] It chelates ionic Fe(III) in the presence of transferrin, but cannot remove iron that is bound to transferrin.[95]

$$NH_2(CH_2)_5\cdot\underset{\substack{|\\ HO}}{N}—\underset{\substack{\|\\ O}}{C}(CH_2)_2\cdot CO\cdot NH(CH_2)_5\cdot\underset{\substack{|\\ HO}}{N}—\underset{\substack{\|\\ O}}{C}(CH_2)_2\cdot CO\cdot NH(CH_2)_5\cdot\underset{\substack{|\\ HO}}{N}—\underset{\substack{\|\\ O}}{C}\cdot CH_3$$

(1)

Ferrioxamines and desferrioxamines are readily soluble in water, but they are poorly absorbed from the digestive tract. This has made desferrioxamine B (Deferoxamine, Desferal) (I) useful, as its methanesulfonate, in the regulation of dietary iron uptake in cases of hemochromatosis.

In idiopathic hemochromatosis (the result of a metabolic anomaly) excessive amounts of iron are absorbed and stored in the liver, spleen, and other organs. Attempts to maintain a low-iron diet, for example by the use of EDTA, can lead to deficiencies of other cations. Desferrioxamine is more nearly specific for iron and, taken orally, it largely blocks the absorption of iron from the alimentary canal.[96] For the depletion of iron deposits within the body, desferrioxamine is given by intramuscular injection or by intravenous drip.[96] Slow addition of desferrioxamine to the bloodstream is desirable if the formation of toxic concentrations of ferrioxamine is to be avoided,[97,98] time being required for ferrioxamine to be excreted through the kidney.

A similar therapy is applied to the treatment of acute iron poisoning, namely dosage with desferrioxamine, orally, to prevent further uptake of iron, coupled with intravenous injection to mask iron already absorbed. Some recent examples of the use of desferrioxamine in the treatment of iron poisoning are given in references.[99–110] There is no doubt that such treatment has greatly improved the prognosis for patients suffering from acute iron poisoning.

Other chelating agents have been suggested for use in this way, notably DTPA,[109,111,112] which in one case was preferred to desferrioxamine on the grounds that it was less toxic.[112] EDTA is also effective, but it leads to the loss of manganese and, particularly, zinc.[113] It may also result in increased absorption of iron.[114] On the other hand, the calcium sodium salts of 2,2′-bis[di(carboxymethyl)amino]diethyl ether and, more especially, 2-(β-aminoethoxy)cyclohexylaminotetraacetic acid, were much better than DTPA or desferrioxamine in preventing the retention of radioiron when they were injected at the same time as $^{59}FeCl_3$.[115]

Injection of desferrioxamine eliminated plutonium from rats more quickly, but less completely, than did DTPA.[116]

SOME GENERAL REMARKS

Masking reactions have been the basis of many suggested methods for overcoming cyanide poisoning. Most of these depend on the dissociation of a cobalt complex, followed by the formation of the stable, nontoxic $Co(CN)_6^{4-}$ ion. Co_2EDTA is one such antidote,[117,118,119,120,121] and it might be followed by treatment with $CaNa_2EDTA$ to chelate any unreacted cobalt ion that is liberated during the treatment.[118] The use of Co_2EDTA is reported to be better than nitrites for this purpose because the latter diminish the hemoglobin concentration by producing the methemoglobin that is needed to react with cyanide ions to form cyanmethemoglobin.[119] Aquocobalamine and cobalt desferrioxamine are claimed to be more effective than Co_2EDTA.[122] Cobalt complexes with histidine[123] and mercaptides such as sodium 2,3-dimercaptopropanesulfamate,[121] and also sodium cobaltinitrite,[124] have been suggested.

Potassium or sodium hexacyanoferate, $M_4Fe(CN)_6$, is not toxic when taken orally, and both were more effective than desferrioxamine, EDTA, DTPA, cyclohexanediaminetetraacetic acid, and several other chelating agents when administered by mouth immediately after injecting toxic amounts of ferrous sulfate into mice.[125] Later treatment by oral administration was ineffective, but intraperitoneal injection diminished the lethal effect.

Effective masking agents have not yet been devised as antidotes or for the removal of many toxic metals, including beryllium, cadmium, manganese and thallium, and this is undoubtedly a fertile area for research.

A suggested future development in the masking of toxic metal ions in the body is the use of lake-forming dyestuffs (such as aurintricarboxylic acid for beryllium) to deposit the metal within the tissues as an insoluble, inert complex, without making any attempt to secure its removal.[126] This approach seems to be potentially hazardous, especially if the organic part of the complex is slowly broken down, with consequent liberation of the metal ion, or if treatment at any time with a more powerful chelating agent could lead to the mobilization of the precipitate.

Finding suitable sequestering agents for the removal of radioisotopes is often difficult, especially when the metal can be deposited in bone or other less readily accessible locations. 2-Mercaptocyclohexylaminodiacetate and its dimethyl derivative were more effective than DTPA in removing ^{210}Pb, but they were too toxic.[127] Bis(dicarboxyaminoethyl) ether was comparable with calcium sodium citrate in removing radioactive strontium from the body, but there was still 50% retention in mice twenty-four hours after simultaneous injection of strontium and chelating agent into different parts of the body.[128] Cyclopentanediaminetetraacetic acid has also been reported to diminish the skeletal retention of radioactive strontium.[129]

From the foregoing discussion some general conclusions can be drawn about requirements for masking agents in the treatment of metal poisoning. The effectiveness of such agents will depend on the following:

1. Their structure and physical chemical properties;
2. The stability of their metal complexes;
3. The distribution of the metals in the tissues and fluids of an organism, and the ways in which they are bound;
4. The ease with which masking agents and their metal complexes can penetrate cell membranes;
5. Their preference for toxic rather than essential cations.

In addition, the following physiological criteria should be met. The masking agent must accomplish the following:

1. Be able to reach the whole vascular system;
2. Be sufficiently nontoxic to avoid complications if overdosage occurs;

3. Prevent the toxic metal ion from penetrating from the bloodstream into cells;
4. Remove the metal ion from those cells into which it has already diffused;
5. Its reaction product with the metal ion must be readily excreted.

13.2 CELLULAR EFFECTS

The growth of microorganisms is often critically dependent on the concentrations of a number of essential metal ions in the nutrient medium. The absence of any one of these metal ions may be sufficient to prevent growth, whereas, if its concentration is too high, the metal ion may be toxic: copper(II) ion is a common example. In culture media, mixtures of metal ions and organic ligands provide buffer systems for maintaining more uniform concentrations of free metal ions, and these concentrations may vary selectively as the ligands are modified or removed by the metabolism of the organism or by pH change.

In earlier biological work, reliance was placed on complexing agents such as citrate and tartrate ions to serve as metal–ion buffers, but more recently emphasis has shifted to synthetic aminopolycarboxylic acids such as EDTA, DTPA, and nitrilotriacetic acid, the metals being added as their chelates directly to the solution. Thus microbiological[130,131,132] and plant tissue[133,134] culture media have been described in which EDTA is used as a metal-chelating and solubilizing agent.

EDTA solutions bring about changes in many kinds of cells: the effects are probably due, in most cases, to the removal of calcium and magnesium ions from cell membranes and cellular fluid. Loss of these cations from the cell wall is believed to be the reason why the resistance of some gram-negative bacteria to antibiotics is lowered by EDTA[135] and why the permeability of the cells is increased.[135,136,137] Similarly, glycine or EDTA sensitizes cells of *Brucella* phage to lysis.[138] Incubation in EDTA solution makes the peripheral regions of ascites sarcoma-37 cells more easily deformable, and this effect is abolished if the cells are reincubated in a saline solution containing calcium ions.[139] Masking of calcium by EDTA suppresses the sticking of leucocytes to inflamed vessels.[140] Similarly, blood platelet aggregation, induced by ADP or collagen, can be prevented by EDTA or EGTA, both of which chelate calcium, whereas addition of calcium or magnesium restores this property.[141]

The ease with which calcium can be displaced from such systems is shown by the fact that 90% of the calcium in human erythrocytes can be removed by washing with EDTA solution, without causing hemolysis.[142] The dynamic nature of equilibria involving calcium and magnesium is seen from the following observations on rat liver microsomes at physiological pH: As EDTA concentration was increased from zero to 20 mM, binding of calcium ion by the microsomes decreased sharply, whereas binding of magnesium increased initially then decreased sharply. With EGTA, similar changes were observed, except that, because the magnesium–EGTA complex is not very stable, magnesium binding to microsomes did not decrease.[143] These results are consistent with expectations from the stability constants of the metal complexes.

Conversely, the depression of oxygen uptake by bone marrow cells treated with EDTA has been attributed to complex formation with cations at the cell surface preventing these cations from activating enzyme systems in the cell interiors.[144]

Sporulation in *Bacillus subtilis* is inhibited by α-picolinic acid, quinaldic acid, 1,10-phenanthroline, and 2,2′-bipyridine because they chelate metal ions that are essential to aconitase activity.[145] Addition of a chelating agent such a EDTA to a culture of an inactivated or attenuated pathogenic agent such as poliomyelitis culture stabilizes the preparation by complexing any heavy metal ions that are present.[146]

13.3 MASKING AGENTS AND ENZYME ACTIVITY

In high concentrations, heavy metal ions such as Ag(I), Cu(II), Hg(II), and Pb(II) inactivate most enzymes and serve as protein precipitants. However, in some cases, even very small concentrations of these ions can lead to loss of activity in highly purified enzyme preparations. Such metal ions might be present in the water used for dialysis or in the buffer solution in which the enzyme activity is measured.

Many examples could be given of metal-ion inhibition of enzyme activity. The following are representative and show how, in general, such inhibitions are reversible if appropriate complexing agents are added to mask the interfering metal ion.

β-Fructofuranosidase is strongly inhibited by Ag(I) which forms a complex with a histidine residue in the enzyme.[147] Magnesium ion inhibits myosin ATP'ase (which is activated by potassium ion), but EDTA restores the activity by complexing the magnesium.[148,149] EDTA, histidine, and 8-hydroxyquinoline protect ATP-creatine phosphotransferase from inactivation by heavy metals.[150] Similarly, oxytetracycline and EDTA prevent this inhibition by copper(II) and nickel(II) ions of sorbitol dehydrogenase from *Acetobacter melanogenus* and *A. suboxydans*.[151] Isocitrate lyase and triosephosphate dyhydrogenase in *Azotobacter vinelandii* extracts are activated by cysteine or EDTA because the latter bind any inhibitory heavy metals present in the assay medium,[152] and

Table 13.1 Some Examples of Enzyme Inhibition by Masking Agents

Enzyme	Source	Comments	Inhibited by
ATP'ase[a]	Blood platelets	Activated by Mg	EDTA
ATP'ase[b]	Guinea pig neocortex	Activated by Ca, Mg	EGTA,[c] EDTA[c]
ATP'ase[d]	Myosin B	Activated by Mg	EGTA
Glucose-6-phosphatase[e,f] / Phosphotransferase[f]	Liver microsomes		1,10-phen, diethyldithiocarbamate, cysteine, azide, NaCN, oxalate
Alkaline phosphatase[g]	Bovine synovial fluid	Activated by Mg	KCN, NaF, EDTA
		Activated by Ca	EGTA
Alkaline phosphatase[h]	Rat intestine	Probably requires Zn	EDTA, NaCN, 1,10-phen, 2,2′-bipy, 8-OH quinoline, pyridine-2,6-diCOOH
Alkaline phosphatase[i]	Serum	Activated by Mg	EDTA
Phospholipase C[j]	*Bacillus cereus*	Requires Zn	EDTA,[c] 1,10-phen[c]
Phospholipase[k]	Rat liver mitochondria	Activated by Ca	EDTA
Deoxyribonuclease[l]	*B. amylozyma*, *Serratia marcescens*	Activated by Mg	EDTA
Nucleases[m]	Growing yeast cells	Activated by Mg	EDTA[c]
Elastase[n]		Activated by Ca	EDTA
Polyphenol oxidase[o]	*Dahlia* tubers	Cu-enzyme	KCN[c]
Cysteamine oxygenase[p]			1,10-phen, diethyldithiocarbamate, KCN, 8-OH quinoline
Dopamine-β-oxidase[q]		Possibly Cu-enzyme	8-OH quinoline, EDTA, 2,9-Me_2-1,10-phen
Alcohol dehydrogenase[r]	Yeast	Zn-enzyme	1,10-phen
Pyruvate oxidase[s]	Heart mitochondria		1,10-phen
Anthranilic hydroxylase[t]	Leaves of *Tecoma stans*	Activated by Fe(III)	2,2′-bipy[c]
Carbonic anhydrase[u]	Bovine erythrocytes	Zn-enzyme	EDTA

[a] B. B. Chernyak, *Vop. Med. Khim.* **13,** 530 (1967).
[b] K. Kadota, S. Mori, and R. Imaizumi, *J. Biochem.* (*Tokyo*) **61,** 424 (1967).
[c] Inhibition is reversible.

[d] A. Okitani, T. Nakamura, and M. Fujimaki, *Agr. Biol. Chem.* (*Tokyo*) **32,** 683 (1968).
[e] R. C. Nordlie and P. T. Johns, *Biochem.* **7,** 1473 (1968).
[f] R. C. Nordlie and P. T. Johns, *Biochem. J.* **104,** 37P (1967).
[g] D. Dadich and O. W. Neuhaus, *J. Biol. Chem.* **241,** 415 (1966).
[h] W. N. Fishman and N. K. Ghosh, *Biochem. J.* **105,** 1163 (1967).
[i] S. Kosmider, *Polish Med. J.* **2,** 338 (1963).
[j] A. C. Ottolenghi, *Biochim. Biophys. Acta* **106,** 510 (1965).
[k] P. Bjoernstad, *J. Lipid Res.* **7,** 612 (1966).
[l] N. V. Sedykh, M. I. Belyaeva, and I. B. Leschchinskaya, *Biofizika* **12,** 546 (1967).
[m] Y. Nakao, S. Y. Lee, H. O. Halvorson, and R. M. Bock, *Biochim. Biophys. Acta* **151,** 114 (1968).
[n] D. A. Hall, *Struct. Function Connective Skeletal Tissue, Proc., St. Andrews, Scotland* **1964,** 116.
[o] Y. Mino, *Obihiro Chikusandaigaku Gakujutsu Kenkyu Hokoku* **4,** 461 (1967).
[p] D. Cavallini, S. Dupre, R. Scandurra, M. T. Graziani, and F. Cotta-Tamusino, *Eur. J. Biochem.* **4,** 209 (1968).
[q] A. L. Green, *Biochim. Biophys. Acta* **81,** 394 (1964).
[r] F. L. Hoch, R. J. P. Williams, and B. L. Vallee, *J. Biol. Chem.* **232,** 453 (1958).
[s] W. C. Yang, D. Yanasugondha, and J. L. Webb, *J. Biol. Chem.* **232,** 659 (1958).
[t] P. Nair and C. S. Vaidyanathan, *Biochim. Biophys. Acta* **110,** 521 (1965).
[u] S. Carpy, *Biochim. Biophys. Acta* **151,** 245 (1968).

γ-glutamyl transpeptidase activity in purified enzyme from lysates of bacteria is increased in the presence of dimercaptopropanol, cysteine, or EDTA.[153] Glutamine synthetase in rat liver is activated by EDTA, presumably by removal of calcium and zinc.[154] With highly purified monoamine oxidase, low concentrations of 1,10-phenanthroline or neocuproine increased the activity of the enzyme, whereas at high concentrations enzyme inhibition took place. A possible explanation of these results is that traces of inhibiting or contaminating metals were removed by low concentrations of ligand, whereas at high concentrations a metal essential to the enzyme activity was masked.[155]

Sometimes, however, metal-ion inhibition is irreversible, particularly in those cases where iron and copper catalyze the aerial oxidation of -SH groups to disulfides. Horse liver catalase preparations are unstable for this reason unless EDTA is added to the aqueous solution to mask the iron present as an impurity.[156] Similarly, in neutral solutions cysteine is protected by EDTA against oxidation by traces of Cu(II).

Conversely, specific metal ions are necessary for the activity of many enzymes. The metal ion may act as such in the solution or it may be incorporated as part of the enzyme itself to form a metalloprotein. In either case masking agents can usually suppress the activity of the enzyme, probably by occupying sites that are necessary for enzyme-substrate interaction. The clotting of blood is a familiar example. Because this is an enzymic process, catalyzed by calcium ion, it is possible to use EDTA, citrate, oxalate, and fluoride ions, or other strong masking agents for calcium as effective anticoagulants. EDTA is more toxic than sodium citrate in transfusion experiments, and blood used for transfusion should contain not more than 0.09% Na_2H_2EDTA.[157] A level of 0.25 ml of 0.1*M* EDTA to 10 ml of fresh whole blood is just sufficient to prevent coagulation by masking the calcium but not influencing the coagulation factors.[158] EDTA (as K_2H_2EDTA) exerts less effect on blood cell counts than does heparin or oxalate ion, and hence appears to be the most reliable anticoagulant.[159]

Again, cyanide ion, and hydrogen sulfide are powerful inhibitors of the respiration of tissues that depend on the iron-containing cytochrome oxidase system because they bind strongly to the metal complexes and inactivate them. Frequently, the dependence of enzyme activity on the presence of a metal ion is shown by dialysis experiments, especially using metal-complexing reagents. When human placental alkaline phosphatase is dialysed against EDTA at pH 5, the enzyme is inactivated unless zinc(II) ion (or, less effectively cobalt(II) ion) is subsequently added.[160] Further examples of inhibition of enzymes by masking agents are given in Table 13.1.

Problems arise in systems containing several different enzymes if a particular metal ion can activate some enzymes and inhibit others, and also in cases where an enzyme requires one kind of metal ion for activation whereas another, and more strongly binding one is present as an inhibitor. There is a great deal of interplay among enzyme systems in a living cell, and deficiencies of essential constituents can affect relative compositions. A selective depression of enzymes has been observed in *Escherichia coli* cells grown in the presence of EDTA.[161] This was attributed to the binding by EDTA of a trace metal, possibly zinc,[162] which was essential for the activity of cyclic phosphodiesterase, 5′-nucleotidase and alkaline phosphates.[161] Phosphodiesterase, present as a contaminant in commercial alkaline phosphatase from *E. coli,* was completely inhibited by adding EDTA.[163] The formation of phosphoserine anhydride by acid hydrolysates of *E. coli* was also inhibited by pretreatment with EDTA, but activity was restored if manganese(II) ion was added.[164]

REFERENCES

1. D. D. Perrin, *Nature* **206,** 170 (1965).
2. D. D. Perrin, and I. G. Sayce, *Talanta* **14,** 833 (1967).
3. See, for example, M. J. Seven, and L. A. Johnson, eds., *Metal-Binding in Medicine,* J. B. Lippincott Co., Philadelphia, 1960; *Biological Aspects of Metal Binding, Federation Proc.* **20,** Suppl 10 (1961); T. N. Pullman, A. R. Lavender, and M. Forland, *Ann. Rev. Med.* **14,** 175 (1963); M. B. Chenoweth, *Clin. Pharmacol. Therap.* **9,** 365 (1968).
4. H. Jacobziner, and H. W. Raybin, *New York J. Med.* **64,** 441 (1964).
5. E. C. Vigliani, and N. Zurio, *Brit. J. Indust. Med.* **8,** 218 (1951).
6. H. M. Perry, and E. F. Perry, *J. Clin. Invest.* **38,** 1452 (1959).
7. S. P. Bessman, H. Reid, and M. Rubin, *M. Ann. District of Columbia* **21,** 312 (1952).
8. F. Rieders, in *Metal-binding in Medicine,* M. J. Seven and L. A. Johnson, eds., J. B. Lippincott Co., Philadelphia, 1960, p. 143.
9. J. Teisinger, I. Prerovsk, and V. Sedivec, *Prac. Lek.* **19,** 251 (1967).
10. P. B. Hammond, A. L. Aronson, and W. C. Olson, *J. Pharmacol. Exp. Ther.* **157,** 196 (1967).
11. H. Mehbod, *J. Am. Med. Assoc.* **201,** 972 (1967).
12. A. L. Aronson, P. B. Hammond, and A. C. Strafuss, *Toxicol. Appl. Pharmacol.* **12,** 337 (1968).
13. A. S. Vakhnitskii, *Gigiena Trude i Prof. Zabolevaniya* **9,** 54 (1965).
14. J. de Bruin, *Nederl. T. Geneesk* **111,** 824 (1967).
15. S. Martin, C. Boudene, R. Truhaut, and C. Albahary, *Arch. Maladies Profess.* **24,** 297 (1963).
16. B. A. Veliev, *Tr. Inst. Kraevoi Patol., Akad. Nauk Kaz. SSR* **10,** 198 (1962).

17. A. Rossi, *Folio Med. (Naples)* **50,** 39 (1967).
18. U.S. Pat. 3,317,378 (1967), W. R. Woods, and K. J. Smith.
19. P. Vohra, and F. H. Kratzer, *Poultry Sci.* **47,** 699 (1968).
20. J. N. Hathcock, C. H. Hill, and G. Matrone, *J. Nutr.* **82,** 106 (1964).
21. P. S. Davis, and D. J. Deller, *Australas. Ann. Med.* **16,** 70 (1967).
22. S. O'Reilly, and W. Bank, *Nature* **212,** 1597 (1966).
23. P. Dervillee, L. Heraut, and E. Dervillee, *14th Intern. Cong. Occupational Health, Madrid* **1963,** 539.
24. V. Eybl, J. Sykora, and F. Mertl, *Acta Biol. Med. Ger.* **17,** 175 (1966).
25. V. Eybl, and J. Sykora, *Acta Biol. Med. Ger.* **16,** 61 (1966).
26. H. A. Schroeder, A. P. Nason, and M. Mitchener, *Am. J. Physiol.* **214,** 796 (1968).
27. W. H. Strain, D. T. Danahy, R. J. O'Reilly, M. R. Thomas, R. M. Wilson, and W. J. Pories, *J. Nucl. Med.* **8,** 110 (1967).
28. F. Havlicke, *Strahlentherapie* **134,** 296 (1967).
29. A. Catsch, *Strahlenschutz Forsch. Praxis* **3,** 182 (1963).
30. H. Kriegel, *Aerztl. Massnahmen Aussergewoehnlicher Strahlenbelastung, Informationstag. Freiburg i. B., 1966 (Oct.), 99.*
31. H. Spencer, and B. Rosoff, *Health Phys.* **11,** 1181 (1965).
32. J. Samachson, J. Dennis, R. Fowler, and A. Schmitz, *Biochim. Biophys. Acta* **148,** 767 (1967).
33. R. A. Gundorova, M. M. Lenkevich, and A. I. Tartakovskaya, *Vestn. Offal'mol.* **80,** 42 (1967).
34. H. Spencer, J. Greenberg, E. Berger, M. Perrone, and D. Laszlo, *J. Lab. Clin. Med.* **47,** 29 (1956).
35. P. Szekely, and N. A. Wynne, *Brit. Heart J.* **25,** 589 (1963).
36. E. Corday, and R. B. T. Skelton, *Amer. Heart J.* **67,** 237 (1964).
37. A. Popovici, C. F. Geschickter, and M. Rubin, *Bull. Georgetown Univ. Med. Center* **5,** 108, (1951).
38. H. Barrie, and B. D. R. Wilson, *J. Am. Med. Assoc.* **180,** 244 (1962).
39. M. Weiner, *Ann. N.Y. Acad. Sci.* **119,** 789 (1964).
40. T. Kochanowicz, *Med. Pracy* **14,** 199 (1963).
41. J. Nosek, and V. Chindar, *Vojenske Zdravot. Listy* **27,** 224 (1958).
42. R. A. Peters, L. A. Stocken, and R. H. S. Thompson, *Nature* **156,** 616 (1945).
43. J. R. Saphir, and R. G. Ney, *J. Am. Med. Ass.* **195,** 782 (1966).
44. L. L. Waters, and C. Stock, *Science* **102,** 661 (1945).
45. H. Eagle, H. J. Magnuson, and R. Fleischman, *J. Clin. Invest.* **25,** 451 (1946).
46. H. Eagle, and H. J. Magnuson, *Am. J. Syph. Gonor. and Ven. Dis.* **30,** 420 (1946).
47. Report of Council, *J. Am. Med. Assoc.* **131,** 824 (1946).
48. A. Cohen, J. Goldman, and A. W. Dubbs, *J. Am. Med. Assoc.* **133,** 749 (1947).
49. C. Ragan, and R. H. Boots, *J. Am. Med. Assoc.* **133,** 752 (1947).
50. W. T. Longcope, and J. A. Luetscher, *J. Clin. Invest.* **25,** 557 (1946).

51. M. Berlin, and R. Rylander, *J. Pharmacol. and Exper. Therap.* **146,** 236 (1964).
52. M. Berlin, L. G. Jerksell, and G. Nordberg, *Acta Pharmacol. (Kobenhavn)* **23,** 312 (1965).
53. L. Magos, *Brit. J. Industr. Med.* **25,** 152 (1968).
54. C. Martinenghi, A. Riccardi, G. Roncari, and V. Morvillo, *Acta Isotop. (Padova)* **6,** 25 (1966).
55. C. Martinenghi, A. Riccardi, and E. Franzetti, *G. Fis. Sanit. Radioprot. Radiaz.* **10,** 208 (1966).
56. K. Lehotsky, and S. Bordas, *Magy. Tud. Akad. Orv. Tud. Oszt. Kozlem.* **18,** 463 (1967).
57. N. Platonow, *Can. Vet. J.* **9**(6), 142 (1968).
58. H. A. Braun, L. M. Lusky, and H. O. Calvery, *J. Pharmacol. and Exp. Ther.* **87** (suppl), 119 (1946).
59. M. D. Stein, and M. A. Perlstein, *J. Dis. Child.* **98,** 80 (1959).
60. O. Grunfeld, *New Engl. J. Med.* **269,** 1138 (1963).
61. J. J. Chisolm, *J. Pediat.* **63,** 843 (1963).
62. J. J. Chisolm, *J. Pediat.* **73,** 1 (1968).
63. R. Coffin, J. L. Phillips, W. I. Staples, and S. Spector, *J. Pediat.* **69,** 198 (1966).
64. A. Gilman, F. S. Philips, R. P. Allen, and E. S. Koelle, *J. Pharmacol. and Exp. Ther.* **87** (suppl), 85 (1946).
65. R. Nath, and S. K. Srivastava, *Indian J. Exp. Biol.* **6**(2), 90 (1968).
66. J. N. Cumings, *Brain* **71,** 410 (1948).
67. J. Marriott, and D. J. Perkins, *Biochim. Biophys. Acta* **117,** 395 (1966).
68. I. E. Okoshnikova, *Prom. Toksikol. i Klinika Prof. Zabolevanii Khim. Etiol. Sb.* **1962,** 205.
69. M. Tati, S. Miyata, and Y. Matsuda, *Igaku to Seibutsugaku* **73,** 35 (1966).
70. Y. Matsuda, M. Tati, and S. Mujata, *Igaku to Seibutsugaku* **76,** 71 (1968).
71. Y. Matsuda, *Gifu Daigaku Igakubu Kiyo* **15,** 869 (1968).
72. M. Tati, S. Miyata, and Y. Matsuda, *Igaku to Seibutsugaku* **73,** 146 (1966).
73. M. Kawai, *Rev. Int. Serv. Sante Armees Terre Mer Air* **39,** 861 (1966).
74. N. I. Luganskii, and Y. I. Loboda, *Farmakol. i Toksikol., Sb.* **1964,** 161.
75. T. Stoichev, *Savremenna Med.* **16,** 356 (1965).
76. D. Soldatovic, and C. Petrovic, *Arh. Farm. (Belgrade)* **17,** 111 (1967).
77. G. F. de Simone, A. Cardinale, and G. Brancato, *Acta Isotop. (Padova)* **7,** 181 (1967).
78. A. Heyndrickx, *Acta Pharmacol. et Toxicol.* **14,** 20 (1958).
79. J. M. Walshe, *Quart. J. Med.* **22,** 483 (1953).
80. J. M. Walshe, *Amer. J. Med.* **21,** 487 (1956).
81. J. E. Wilson, and V. du Vigneaud, *J. Biol. Chem.* **184,** 63 (1950).
82. E. J. Kuchinskas, and V. du Vigneaud, *Arch. Biochem. Biophys.* **66,** 1 (1957).
83. See, for example, J. Lange, *Deutsch, Med. Wschr.* **92,** 1657 (1967).
84. I. Sternlieb, and I. H. Scheinberg, *New Engl. J. Med.* **278,** 352 (1968).

85. T. C. Chalmers, *New Engl. J. Med.* **278,** 910 (1968).
86. J. M. Walshe, *New Engl. J. Med.* **278,** 795 (1968).
87. T. J. Fagan, *New Engl. J. Med.* **278,** 1124 (1968).
88. J. E. Boulding, and R. A. Baker, *Lancet* **ii,** 985 (1957).
89. K. Cramer, and S. Selander, *Postgrad. Med. J., Oct. 1968* (suppl), 45, who also review earlier literature on the use of penicillamine for removing lead.
90. V. Parameshvara, *Brit. J. Indust. Med.* **24,** 73 (1967).
91. H. Bickel, G. Hall, W. Schierlein, V. Prelog, E. Vischer, and A. Wettstein, *Helv. Chim. Acta* **43,** 2129 (1960).
92. V. Prelog, and A. Walser, *Helv. Chim. Acta* **45,** 631 (1962).
93. G. Anderegg, F. L'Eplattenier, and G. Schwarzenbach, *Helv. Chim. Acta* **46,** 1400 (1963).
94. H. Keberle, *Ann. N.Y. Acad. Sci.* **119,** 758 (1964).
95. S. P. Balcerzak, W. N. Jensen, and S. Pollack, *Scand. J. Haemat.* **3,** 205 (1966).
96. S. J. Hopkins, *Pharm. J.* **139,** 363 (1964). See also J. L. Fahey, C. A. Rath, J. V. Princiotto, I. B. Brick, and M. Rubin, *J. Lab. Clin. Med.* **57,** 436 (1961); B. Karlsson, R. Lagercrantz, and I. Braun, *Nord. Med.* **74,** 985 (1965).
97. C. F. Whitten, G. W. Gibson, B. S. Good, J. F. Goodwin, and J. A. Brough, *Pediatrics* **36,** 322 (1965).
98. C. F. Whitten, Y. Chen, and G. W. Gibson, *Pediatrics* **38,** 103 (1966).
99. R. M. Bannerman, S. T. Callender, and D. L. Williams, *Brit. Med. J.* **1962** (*ii*) 1575.
100. W. F. Westlin, *Clin. Pediat.* **5,** 531 (1966).
101. F. Henderson, T. J. Vietti, and E. B. Brown, *J. Am. Med. Ass.* **186,** 1139 (1963).
102. S. Moeschlin, and U. Schnider, *New Engl. J. Med.* **269,** 57 (1963).
103. J. T. McEnery, and R. B. Mack, *Illinois Med. J.* **126,** 550 (1964).
104. N. Shapiro, and G. O. Barbezat, *S. Afr. Med. J.* **38,** 461 (1964).
105. A. Santos, and A. Pisciotta, *Am. J. Dis. Child.* **107,** 424 (1964).
106. J. Jacobs, H. Greene, and B. R. Gendel, *New Engl. J. Med.* **273,** 1124 (1965).
107. J. T. McEnery, and J. Greengard, *J. Pediat.* **68,** 773 (1966).
108. S. Leikin, P. Vossough, and F. Machir-Fatemi, *J. Pediat.* **71,** 425 (1967).
109. D. G. Barr, and D. K. B. Fraser, *Brit. Med. J.* **1968** (**i**) 737.
110. J. T. McEnery, *J. Pediat.* **72,** 147 (1968).
111. T. J. Covey, *J. Pediat.* **64,** 218 (1964).
112. G. Pfister, A. Catsch, and V. Nitrovic, *Arzneim.-Forsch.* **17,** 748 (1967).
113. J. Tripod, *Atti Accad. Med. Lombarda, Suppl.* **20,** 2057 (1965).
114. W. R. Bronson, and T. R. C. Sisson, *Am. J. Dis. Child.* **99,** 18 (1960).
115. L. du Khuong, *Arzneim.-Forsch.* **15,** 387 (1965).
116. R. Truhaut, C. Boudene, M. Lutz, and H. Metivier, *Arch. Mal. Prof., Med. Trav. Secur. Soc.* **27,** 669 (1966).
117. M. Terzic, and M. Milosevic, *Therapie* **18,** 55 (1963).
118. H. Wehrheim, and K. Jacobs, *Z. Arbeitsmed. Arbeitsschutz.* **15**(5), 107 (1965).

119. E. L. Knowles, and J. T. B. Bain, *Chem. Ind. (London)* **1968,** 232.
120. G. Paulet, *Arch. Maladies Profess.* **22,** 120 (1961).
121. O. E. Kolesov, and V. N. Cherepanova, *Farmakol. i Toksikol. Sb.* **1964,** 167.
122. K. D. Friedberg, J. Gruetzmacher, and L. Lendle, *Arch. Toxikol.* **22,** 176 (1966).
123. H. Mercker, and G. Bastian, *Arch. Exp. Path. Pharmakol.* **236,** 449 (1959).
124. R. M. Worth, K. Kikuchi, and K. K. Chen, *Proc. Soc. Exp. Biol. Med.* **120,** 780 (1965).
125. V. Nigrovic, and A. Catsch, *Arch. Exp. Path. Pharmakol.* **251,** 225 (1965).
126. J. Schubert, *Federation Proc.* **20,** (Suppl. 10), 203 (1961).
127. A. Catsch, *Arzneim.-Forsch.* **17,** 493 (1967).
128. A. Kubodera, T. Mori, and S. Tsurufuji, *J. Pharm. Soc. Jap.* **87,** 511 (1967).
129. K. Kostial, S. Vojvodic, and T. Maljkovic, *Arh. Hig. Rada Toksikol.* **18,** 111 (1967).
130. S. H. Hutner, L. Provasoli, A. Schatz, and C. P. Haskins, *Proc. Am. Phil. Soc.* **94,** 152 (1950).
131. S. H. Hutner, and L. Provasoli, in A. Lwoff, ed., *Biochemistry and Physiology of Protozoa,* Academic Press, New York, 1951. Vol. 1, p. 29.
132. L. Provasoli, J. J. McLaughlin, and M. R. Droop, *Arkiv. Mikrol.* **25,** 392 (1957).
133. J. Myers, J. N. Phillips, and J. R. Graham, *Plant Physiol.* **26,** 539 (1951).
134. R. M. Klein, and G. E. Manos, *Ann. N.Y. Acad. Sci.* **88,** 416 (1960).
135. R. Weiser, A. W. Asscher, and J. Wimpenny, *Nature* **219,** 1365 (1968).
136. J. M. T. Hamilton-Miller, *Biochem. J.* **100,** 675 (1966).
137. L. Leive, *J. Biol. Chem.* **243,** 2373 (1968).
138. L. M. Jones, G. S. Merz, and J. B. Wilson, *Experientia* **24,** 20 (1968).
139. L. Weiss, *J. Cell. Biol.* **33,** 341 (1967).
140. P. L. Thompson, J. M. Papadimitriou, and M. N. I. Walters, *J. Path. Bact.* **94,** 389 (1967).
141. T. Hovig, *Thromb. Diath. Haemorrhag.* **12,** 179 (1964).
142. D. G. Harrison, and A. B. Sidle, *Biochem. J.* **108,** 40P (1968).
143. H. Sanui, and N. Pace, *J. Cell. Physiol.* **69,** 11 (1967).
144. R. M. Gesinski, and J. H. Morrison, *Experientia* **24,** 296 (1968).
145. P. Fortnagel, and E. Freese, *J. Biol. Chem.* **243,** 5289 (1968).
146. French Pat. 1,403,608 (1965), Behringwerke, A. G.
147. K. Myrback, *Arkiv Kemi* **11,** 47 (1957).
148. A. Muhlrad, F. Fabian, and N. A. Biro, *Biochim. Biophys. Acta* **89,** 186 (1964).
149. G. W. Offer, *Biochim. Biophys. Acta* **89,** 566 (1964).
150. W. J. O'Sullivan, and J. F. Morrison, *Biochim. Biophys. Acta* **77,** 142 (1963).
151. J. Huber, T. Elsaesser, and H. Hilscher, *Z. Allgem. Mikrobiol.* **3,** 136 (1963).
152. I. R. Kennedy, and M. J. Dilworth, *Biochim. Biophys. Acta* **67,** 226 (1963).
153. A. Szewczuk, and M. Mulczyk, *Arch. Immunol. Ther. Exp.* **15,** 395 (1967).

154. L. Hsu, and A. L. Tappel, *J. Cellular Comp. Physiol.* **64,** 265 (1964).
155. S. Gabay, and A. J. Valcourt, *Biochim. Biophys. Acta* **159,** 440 (1968).
156. French Pat. 1,489,707 (1967), Laboratoires Sarget-Ambrine.
157. E. Skala, *Nouv. Rev. Fr. Hematol.* **8,** 284 (1968).
158. E. Goettinger, and H. Vinazzer, *Thromb. Diath. Haemorrhag.* **11,** 513 (1964).
159. S. Miwa, K. Teramura, and I. Takahasi, *Rinsho Byori* **15,** 376 (1967).
160. D. R. Harkness, *Arch. Biochem. Biophys.* **126,** 513 (1968).
161. H. F. Dvorak, *J. Biol. Chem.* **243,** 2640 (1968).
162. H. F. Dvorak, and L. A. Heppel, *J. Biol. Chem.* **243,** 2647 (1968).
163. W. M. Lewis, and M. E. Hodes, *Life Sci.* (*Oxford*) **7,** 251 (1968).
164. G. W. Rafter, *Arch. Biochem. Biophys.* **122,** 648 (1967).

INDEX

DATE DUE

7/13/89	SLS		